TABLE OF CONTENTS

Chapter One: The Mystery of What's Beyond A Cell
Pg. 1

Chapter Two: Recognizing Cognition: Past and Present:
Pg. 29

Chapter Three: Light
Pg. 66

Chapter Four: The Purpose of Humanity
Pg. 83

Chapter Five: The Battle Between Good and Evil
Pg. 110

Chapter Six: Evolution's Aspiration
Pg. 128

CHAPTER ONE: THE MYSTERY OF WHAT'S BEYOND A CELL

Getting The Question Right

What is the purpose of life? This question has dominated the human equation probably more than any other question in the history of man. And yet the question itself is flawed because it presupposes that we have a strong foundational understanding for the thing we know as: "life." Currently, when we ask the question what is the purpose of life, we are usually (rightly or wrongly)) focusing upon the purpose of human life. Why? Because when we attempt to observe life we see it most predominantly first through the eyes of ourselves, and then second, through the eyes of our fellow man. And when that's done, we then tend to compare and contrast ourselves to the other species of organic life (native to our world) such as all forms of plants, sea life, insects, animals, reptiles and mammals, and … so on. So when we ask the question of what is the purpose of life; we tend to look for the answer first with the assistance of a comparative and relativistic analysis. Why? Because the purpose of life within "humanity;" is both internal to each individual and relativistic to all the other humans; both past and present. So when we ask the question what is the purpose of life, we have to break it down into its component parts. First we have to allow each individual the benefit of being personally introspective; by allowing for their first natural, almost instinctive response of questioning: what is the purpose of "their life?" Then we must allow the enquirer to assess their personal life as a thing in transition, changing through time. The perceived purpose of a twenty year old, is likely to be significantly different then when that same person reaches fifty. Then we must allow that same observer to reflect upon their life in comparison to all other humans, with an equal responsibility to reflect upon humanity as humanity itself has changed or evolved over time. And then, once all those distinctions are at hand, we can then ask the broader general question of what is the purpose of "human life" in conjunction with/ and in contrast to, all the other forms of life known to us; because that contrast allows us to reframe the question of what is the purpose of life, to the broadest possible scale imaginable, meaning that we can then contemplate what is the purpose of all life, writ large.

Now of course human life, as a collective species, is readily recognizable as being only a single sample of the many forms of life here, on our planet. And again of course it may be natural or instinctive to then look for some meaning to our lives and to our search for the purpose of "life", to use some method of comparative analysis with respect to those other forms of (known) organic life to give our search (for the meaning of life generally) some baring or at least some sort of comparative point of reference. But again, such an attempt may be premature if we do not have a foundational understanding for the concept of "life" to begin with. For example, what if it is premature, or just simply wrong; to assume that life is only to be understood from the singular perspective of organic life. What if life does exist past (organic) death? What if life continually exists between lives? And, are memories irrelevant? What if life exists as a continuum? And what if such a transitional continuum, can not only be contemplated; but (if we actually looked) could be proven, tested, measured and quantified?

So again, back to the beginning; before we can give some understanding to the "purpose" (or meaning) of "life;" we may have to have a more profound (foundational) construct for what it is that we understand to be "life." And on this, it is probably best that we at least start from the premise that we do not know what "life" (writ large) actually is; we can only surmise that our search for an understanding for all life will be a search that starts somewhere in the middle of time, space, evolution and intellect. Where our understanding of life will be premised positively on the sophistication, knowledge and observations that we currently have; while realizing that our best efforts to understand "life," will unfortunately be undermined by: what we don't know, what we can't see, and what we can't, or haven't (as yet) even imagined.

And so at this point it is also important that before we now take on this universal challenge of comprehending this thing that we call "life;" that we now pause to remember that our own analysis is to be forever compromised by the fact that: it is "we humans," who are in fact doing the analysis. So one question which will always persist for us, is: how objective can/or will, our ongoing analysis ever actually be?

And before we can actually leave this area of concern, there is one other question that is worthy of merit and that is the question of (when the generic term "we" is used) who is it that is being referred to as "we?" Or in other words, when a generic reference to humanity is being made through the use of the term "we;" we need to know of "whom" it is, of which we speak. Why? Because the social and societal accomplishments of mankind to date, have still left us with a variety of societal matrices (worldwide): where the pursuit of power and material gain, coupled with the vices of ego, arrogance and all other forms of human failings; can corrupt our most honest inquiries into our very knowledge of life. Consequently, we must therefore realize from the outset that (as in all things) reliable factual knowledge can be corrupted through hidden and ulterior (human) motives both in the chosen method of our pursuit, and through to the integrity of the conclusions that we ultimately choose to draw. Quite simply, the old axiom is therefore true;: knowledge, is power. And human consciousness is therefore corruptible in any number of multi dimensional ways. So if we are to understand the science of life, we must proceed from the stand point that all of our analysis must be premised on the purity of our facts. Similarly, our extrapolated conclusions must also be solely determined by the pure logic that must necessarily follow. In science, therefore; the concept of "knowledge" must be by definition, unencumbered (both in substance and in process) by any of our many human frailties; true knowledge therefore must not be contaminated by the poisons of our mind and our intellect, such as ego, pride, arrogance, selfishness, greed, ambition, vanity; and … unfortunately, this is to name just a few.

Despite Ourselves, The Best of Us Have to Try

That said, however, if we are to do what we can; if we start our journey for understanding "life" given the constraints and the biases alluded to thus far; then our initial definition of life; must be understood to be vastly incomplete because (even with respect to the positive) it will be: limited to, and again compromised by, the level of our knowledge and our sophistication that we do have; and constrained (in the negative) by our biases and prejudices that any particular "we" may have at the time that the question is being asked.

That said however, we can guard against such limitations and vices, if we accept from the outset that our search for an understanding of "life" can and should be fortified by a determination to purge our analysis through the test of (and by contrast to) the results found from a multiplicity of diverse disciplines. This is to say that where the frailties of any one perspective can compromise the veracity of any one pursuit, true knowledge can become more reliable when the conclusions reached in one field are buttressed and found to be compatible with other conclusions from other independent and divergent sources. So our analysis for the pursuit of "life," be it with regards to its "origins," "meaning or purpose;" the overall integrity of such an analysis will best be protected and advanced reliably through an ever deepening appreciation and sophistication for a compatible understanding (dare it be said) for "all things." Why, because knowledge through multiplicity of disciplines, will come with its own (unrelated and objective) verification, and the strength of confidence (earned through humility) as we learn to accept: …all those things that we don't actually know, … about everything.

From Microbiology…And Beyond.

For example, in our efforts to understand all life (at its broadest), does it not make sense that we include in our approach to understanding life; that we (through humility) include the possibility that anything and everything; could have an impact on bringing about life, even those things that we may currently scientifically believe to be dormant; or contrary to; or something that is itself held to be lifeless. Such lifeless, or contrarian things, need not be meaningless in understanding life; if through such other things, such as contact, or by design, or through the introduction of new formulae, these lifeless things (in ways we do not as yet understand) are themselves some sort of conveyor, catalyst or formulae for a transformative development in the things we do hold to be life forms.

For instance, currently, there is an academic debate whether viruses exist as life forms. Clearly, viruses are not life forms in the same sense that we consider cells to be life forms. It is argued that: if viruses are life forms at all, they are to be considered to be an entirely different form of life because they cannot exist on their own. Their existence consists of (at least) a nucleic acid surrounded by a protein coat, and a sequencing program of DNA origin. Originally however, before 1931, viruses were thought to be microscopic cells, but in 1931 with the advent of the electron microscope, viruses were rethought to be more regarded as genetic aberrations because: all viruses need a host to survive. A virus cannot replicate or metabolize by itself; and they can only replicate (once inside a cell) by infiltrating (and hijacking) the cellular structure and chemicals of its host. Thus with such distinctions, viruses (again if viewed as a life form) can only be argued to be a (hostile) life form; differentiated from cellular life, because that which is missing in itself, must be acquired (or stolen) to become alive, but once it has acquired what it needs, it replicates so quickly that it becomes itself, a self destructive contradiction. Self destructive, because either the host's immune system rallies its defenses (through white blood cells) and kills off the virus, or; the virus itself kills off sufficient numbers of the host's cells, to kill off the host; leaving the virus itself to then subsequently die.

Now why is this so very important, it is important because the virus paradigm described above, demonstrates that if a thing, such as a virus is not a life form because it exists (in part) as dormant code while it is outside any life sustaining host. Such physical matter as this is therefore lifeless by definition because it is (in part) code, without the chemicals or the independent structure to manifest itself; thereby existing only as a speculation of something that might be. Or (paradoxically) it is a life form that exists as a different kind of life form, as it is incomplete in missing some essential elements, but has a (construct) code that will enable it to become alive if sustained by/ and with, the sacrifices of its host.

The point being made here is that (by human definition); either a virus is a different kind of life form (adverse to the existence of cellular life), that exists in part as a code; or it isn't a life form at all, but supposedly a lifeless thing that has the ability to bring itself (with a dormant code) to life. And this of course, creates the possibility that (at the very least) that which is dormant/ or partly dormant, can recognizably (through a dormant code) become/or contribute to (or defeat) life in ways that (we at any given time) might be unable to understand. So to boiler plate this example further, either a virus is a life form, but different, because it is in part dormant code; or it is not a life form because it is part dormant; but either way that part which makes it either a life form (but different), or lifeless; either way both characterizations recognize that that which lies dormant as code can (when properly stimulated) become life. And that means that (if not in whole/ but in part) that which appears to be without life can sometimes create life. The problem to date; is, that no life (as we understand the term) has been found to exist (as yet) that originated simply from code alone.

So if we are to deduct from this that we should be prepared to look for the origins of life everywhere and anywhere, even to understand that life might possibly "reside"/ dormant or not; as opposed to actually exist (in a living form) by residing dormant where we might least expect it, then doesn't it become logically possible that life could exist in an unrecognizable form, including in a dormant state (maybe even existing in a state of stasis) as a formula or code here on earth. A formula or code, that just needs to be activated; through some sort of stimuli, for itself, to become alive. So, to be clear; genetic code generally exists as recordings to be reproduced purely as reflections of a thing past. But in those instances where the codes were originally generated by matter with "life harboring" potential, then in those instances, a potential life form with a dormant residual code might be the next evolutionary step in the adaptation for an (as yet) unrecognizable strain of life. And on this very point: the evolution of our own human DNA may be just demonstrative of this possibility.

For instance, human DNA as we understand it today is unclear how it actually came about. Current theorems hypothesize that various forms of cellular DNA are themselves descendents of a previous sequencing precursor within cellular structure called RNA's which had either a weaker or less efficient genome for cellular repair; the argument being (in this much simplified overgeneralization) is that DNA sequencing (in its many different forms) surpassed RNA sequencing (through natural selection) because within humans, DNA sequencing simply had a more efficient genome for cellular repair, than RNA. And that (through time and further evolution) DNA sequencing replaced RNA as an alternative sequencing method by simply allowing RNA (for humans) to die off; again, because it had a much less efficient sequencing precursor for cellular adaptation and survival. Alternatively however, other theoretical scenarios hypothesize that the more efficient genome for our cellular repair possibly originated in the adoption of DNA sequencing as various cells came in contact with "viral DNA" that had the stronger genome for repair; and that cellular organisms (in order to simply survive), adopted (or through invasion) had a viral genome imposed on their cellular DNA sequence (because again; it had the more resilient genome for structural repair).

And the important point to be made here, is that: by either means of creation (selection or invasion), our current DNA sequencing code (as we have the benefit of it today), this very important building block for all cellular life (with its current matrix of genomes) may have come from a foreign (non cellular) code that may have been (and may still be) incapable of surviving on its own, but as code it can be capable of remaining dormant and lifeless until such a code could/or can, become beneficial to other living cells. And this may have come about either through natural cellular selection and adoption; or through cellular victory, after a foreign (viral) entity's invasion.

So from the above example, our comprehension for the potential of the possibility of contributing code (viral or otherwise) causing a structural change in cellular life, such a possibility as this logically mandates that any honest search for an understanding of life (at its broadest) is that we must be prepared to pursue, and be open to all that which may be discovered from the most expansive possible construct that we can imagine. And this means that where we look upon life primarily from an organic perspective, where our biased search for the meaning of life (as postulated earlier) transcends from our own initial inquiry as to the meaning of our own individual human life, to an almost infinite transcendence towards the subatomic. That is, as mentioned earlier, there is a recognizable psychological (personal) progression as each individual ponders the meaning and purpose of life, it is understandable that each of us might first personalize the question as it might immediately apply (i) to them directly; but once one moves past the specifics of an individual, the next transcendence requires (ii) an examination of humanity as a collective with a regard to the evolution of us as a society through history. Then, from there, what follows is: the transcendent analysis of mankind as we trace events and choices made past the known history of mankind; we would then transcend to (iii) an understanding for the evolution of man as a species. Next, to then (for perspective): (iv) contrasting our species from the other divergent (evolving) species created through the science of biochemistry; to then ultimately (v) exploring the molecular science that has established that all life on earth is premised on the carbon molecule. From this transcendence we would then be able to see that our search for life, and our (more personal and collective) pursuit for the meaning of life; must (because of our findings) comprehend the possibility that life as we know it is of a collaboration of our earth's "current life." That is, a collaboration, where life in each of its consecutive earlier stages has more of a common denominator root of origin than the vast diversification of life that it has now. Or to say it more succinctly, as the foundational building blocks for life of each species become more repetitive the further back in evolution that we go, the more logical it becomes that the purpose of life might be more similar than what the current diversification of life in its multitude of species might suggest. That is, that as we look backward in time, we will find, that once we get closer to the beginning, the diversification of species life actually serves a more common/sole purpose in the broadest possible context. That is, that we may realize that life itself; has (through our understanding of all the species) for us a common purpose (beyond the immediate need to survive) relevance in contributing to help us answer the overall question of: what is the purpose of life.

So, if we go back to the beginning; the fact that our analysis starts from a very personalized and individualistic perspective, highly dependent on the individual psychology of each individual, the transcendent process for all life backwards through time and evolution points more and more to a more collective common foundational origin. And this common origin creates the possibility that despite a vast array of diversification, divergent species may still be acting as a whole; despite the fact that each species seems to appear to evolve individually. Meaning that at some previous evolutionary divergence there is imprinted a compulsion that each species may still be working to contribute to the evolution of the whole (but each differently) as each works in collaboration with a greater collective, through the evolutionary process of trial and error. And this possibility means that we have to be open to the possibility that there is either (some sort of underlying or over riding) directional cognitive property to do so. Or to put it differently, if each species is analogous to a worker on an assembly line, and when a particular species figures out how to make a better / or tighter screw, that worker/ species in endeavoring to make that improvement, would need an overseeing floor manager that would recognize that alteration (and all other species improvements) for their contribution, and then consciously synthesize that alteration into the final design and creation of an end product. But in an assembly line analogy where the end goal is typically to create more (numerous) mass production; conversely however, in the case of earthly evolution, the efforts of the many (species) might be to create a more perfect "life" as a harmonious whole. And on this, we would do well to remember that if treated as a whole, the earth as a thriving entity might best be regarded collectively, as a single entity, where all the advances that take place after the achievement of a collaborative consciousness can be utilized (based on the level of sophistication of that consciousness) for the benefit of the whole. And, not surprisingly, while all species are divergent, until a heightened form of communication between all species comes along; all of these advances would/and will have to be assembled consciously through the conscious watchful eye of …observation alone. That is: that in the absence of an understood dialogue, all communication between species is best understood through the process of "watch and learn." And the larger question is, do we have the humility and wisdom to realize what there is to be learned.

Flight: It's Been There, Right In Front of Us, For Thousands of Years.

And on this point the human ability to conquer the ability of flight as an evolutionary step is directly on point. For it has been known by man by watching birds for millennium, that wings were critical to flight as they create the necessary lift, but what we have struggled with (in man's aspiration for flight) is first: the creation of propulsion necessary for the weight/to lift ratio needed to give man flight; which was achieved through the advent of the combustible engine and the propeller. And second: it was the astute observation by the Wright brothers to discover the mechanics of aeronautic control; through wing twists and flap manipulation that gave mankind control within flight that allowed man-made flight to become possible. And this took place through the observation that birds steer by virtue of pivoting the directional wing downwards in the direction they wish to turn, while twisting the alternate wing upwards creating an opposite correlation within the two wings, where the two wings work in conjunction with one another.

So the earth working as a singular entity created the necessary component parts to allow certain species to demonstrate what is physically possible (namely in this case, flight), while simultaneously (in cosmic relativity) the human species were developing that consciousness which could watch, learn, correlate, innovate and adapt to take everything to the next level. So regardless of what flattering things we may think intellectually of ourselves; this is clearly what the earth has accomplished, especially when its component parts work consciously together.

So the question then becomes, does a driving force for life to evolve and aspire come from earthly matter by chance; or is there a directional collaborative driving force, that exists with/ or without form, that puts our evolutionary course on the trajectory that has brought us to where we are now. Either way, if our transcendent understanding is to (ever) be complete, our understanding of life must include any foundational contributions made to life, both from the divergent species level; but also, despite this divergence, we will need to also allow for a transcendent understanding for any discovery of binding commonality that might exist at either the celestial, or the atomic (and/or subatomic) levels.

The Subatomic Unknown

So from the physical multidimensional world of which we know, not only is it necessary for our search for an understanding of life to cause us to look outward to the cosmos but equally we must be prepared to simultaneously pursue an infinite, multidimensional (internal) transcendence deep into the subatomic. Why, because although subatomic theory does not typically correlate particle motion (currently) with the origins of life; if we are to expand our definition and search for life to include all that may have life generating (or compatibility) properties; such forces that are consistent with energy particles that have (self contained) motion held within (such as the electron movement internal to an atom); then clearly such transcendence will require our appreciation to include a deeper understanding (of at least) electron particle motion. This is to say that what is currently unknown about the nuclear realities within the subatomic universe with all of its immense powers; it is these unknowns which forces us to contemplate that (supposedly lifeless) matter through its bond with particle motion may hold as yet undisclosed properties for the propulsion of life. For instance, with respect to the phenomena of ascertaining subatomic energy and motion, we now know that electrons within an atom, do not actually orbit a nucleus, as was once originally represented. In the Rutherford (symbolic) schematic, dots would circle a representative nucleus core in a seemingly linear two dimensional orbit; but now we know that such a representation of an electron's motion is no longer accurate, rather the more accurate representation is that electrons are now proven to exist more as three (or multidimensional) spheres that can not only surround a core; but that these spheres are also capable of assuming different spherical shapes and trajectories. Thus if these more current schematic models prove themselves to be accurate, then is it not also possible to characterize such electron spheres as being similar to, some sort of electron based/electromagnetic "membrane." Even though, we have never chosen to use the word membrane in this fashion before. Moreover, previous descriptions of an atoms existence explained that given the absolute subatomic/microscopic size of an atom's particles namely: protons, neutrons and electrons, that the space in between these particles in contrast to their size, left the space between the electrons and their nucleus as 99.999999 % potentially open space.

But we now know that this too, this open space characterization is also flawed; as well. Yes: it is still scientifically maintained that the space between the nucleus and the electron spheres are inexplicably immense, in contrast to the size of the particles themselves; but the notion that such spaces are empty or void of any "stuff," is no longer theorized. Alternatively, what has been postulated is that these areas can be described as orbital clouds, and that within these domains can be found quantum fields, electromagnetic fields, virtual particles, photons and other phenomenon that represent energy variations which are described (in the abstract) as "excited probabilities." So although there remains inexplicable space enough to be further explored, the original premise that there were spatial voids of emptiness between the particles is now known to be incorrect; and therefore the probable existence of other forms of (currently incomprehensible) theoretical phenomenon, must now not only be (both logically) anticipated; but now, must be theoretically considered for all their infinite quantum possibilities. Furthermore this conclusion that life's most infinitesimally small propelling properties might be found anywhere, is further supported by the recent findings that as we progress ever further through our transcendence deeper into the subatomic, we have now confidently established that within an atom's nucleus, protons can be further broken down into further component parts of hadrons and quarks (and other things); and that there is in fact amongst these particles a verifiable binding force that compels these other sub nuclei realities to remain bonded within; and yet this binding "strong force;" is an inexplicable force, of which we know almost nothing about.

And although this conclusion appears to leave us again: back at the beginning, it does not; because now the question (from our current middle position of evolutionary consciousness) recognizes the possibility that all these things are all interconnected and that the existence and advancement of all things may act as catalysts and conveyors for one another; and this creates the possibility that everything is part of a larger equation. Therefore it follows that in any search for an understanding of life; all things need to be included, including all things that may even appear to be innate or lifeless even at the subatomic level. And it further means that wherever there is bonding of any imaginable particles or entities, as our worldly realities intersect, the potential for the discovery of dormant or inexplicable influences for the "original building" of all matter; such potential, therefore cannot be ruled out. So to be blunt, where there is motion of any kind, whether that motion be, past or present (past in the sense of things that now exist only as code); there may still exist realities (possibly dormant) which correspond with forces therein, which may have contributed to the "origins of all things"; which may have through time and progression, inevitably lead to the "origins of life."

Realities Without Form

But of equal importance, in our search for life, from the subatomic to the celestial; must also be our willingness to allow our intellect to grasp that which goes beyond the physical, beyond the multi dimensional; we have to allow ourselves to comprehend the existence of the "empyreal" as if it were as real and as consequential, as all other things.

For example, "mathematics" as we currently understand it is itself either a thing or it is a representation of a thing that is itself, "empyreal." Mathematics is perceived as a thing because it exists as a language that allows us to conceptualize, describe, measure, quantify and discern all that we see allowing us to comprehend objectively that which we perceive as physical reality. We perceive mathematics as a thing because it not only allows us an objective scale upon which we can comprehend that which is around us, but it also allows us to predict and measure movement of matter with accuracy and precision. We know mathematics is representative of something because in motion we see our mathematical predictions hit their calculated targets. Mathematics is real in this sense because we did empirically and successfully land a drone on Mars. But mathematics is still however itself (from a human perspective) a conceptual reality. Mathematics in its most independent reality (meaning beyond the conceptualization of man) exists most as a thing because apart from man's conceptualization of numbers, math has its own independent reality through the undeniable existence of "symmetry." In fact, mathematics as a thing is just a description of "symmetry." Symmetry is what math is all about. Symmetry is math. And symmetry, like "perfect harmony" is real because such things … are just undeniable. On this the old proverb is applicable: "if a tree falls in the woods, but there's no one to hear it, did it make a sound?" And the unequivocal answer is: yes; because symmetry and math require no observer to exist, they just simply, "are." The vibrations we identify as sound are just one type of the possible consequence of matter movement and collision, so in our parable, because "the tree fell," what followed … just simply followed. And how we identify with what actually followed; … just doesn't (really) matter. Put simply: that which "is;" doesn't need our acknowledgement nor our comprehension, to actually … be. Even back in the 1500's, (from an entirely different discipline): "A rose by any other name, would smell as sweet." Even if there was no physical ability to smell at all.

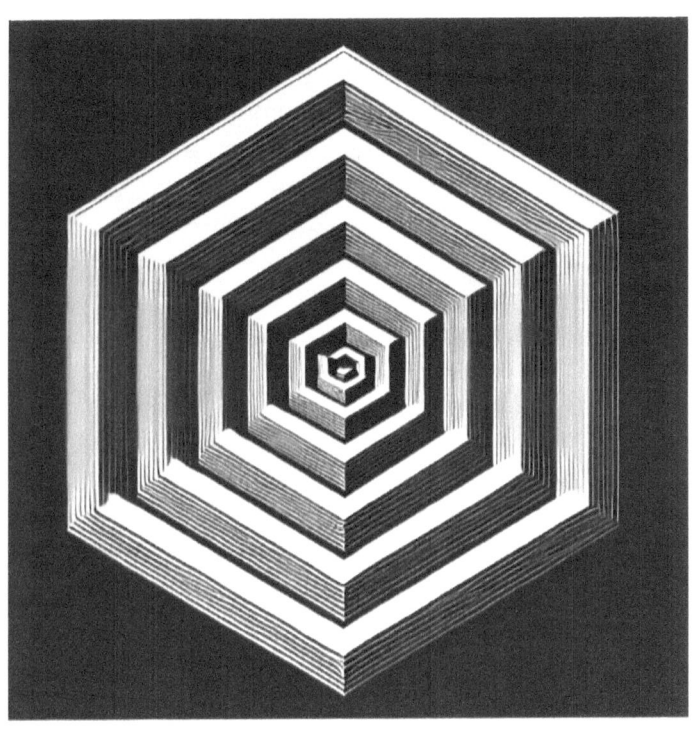

A Hexagon: Or An Infinity Cube.

So the conclusions that can be drawn from the above is that: a particular thing can exist in three dimensions (or more) and therefore can be easily regarded as real, but also an abstract conceptual (empyreal) thing, such as symmetry, can also be real; even though it has no form. Geometry is real: but it cannot be weighed or measured, geometry exists like symmetry, because its proof of existence is in its "precision and accuracy," so in many ways the" truth" held within geometry can be considered a subcategory of symmetry. ... So if we can acknowledge that in all things there is the potential for both the tangible and the empyreal, why should our analysis of life (and its meaning) be any less sophisticated or multidimensional?

So from the above, the logical choice for a search for an understanding of life is that we must be prepared to pursue that which may be discovered from the broadest possible (multidimensional) intellectual construct that we can imagine. So from the usual three dimensional world of which we know, it is necessary for us to simultaneously pursue the infinite transcendence into the subatomic, while also looking outwards to the furthest reaches of our understanding for our cosmos. But of equal importance, there must be our willingness to also allow our intellect to grasp that which goes beyond the physical; that is: we have to allow ourselves to comprehend the existence of the "empyreal" as if it were as tangible, as all other usual things. For instance, as a question; is it within our realm of comprehension (once we can visualize and accept the independent existence of such things as "symmetry" as being an actual thing with a reality unto itself) to then ask the next logical question of whether such unknown empyreal things, even things without form (such as symmetry) could such things exist, that then have had an impact on the "origins of life."

So if we are brazen enough to ponder the origins of life, we must (through humility) also query whether we are evolved enough to conceptualize that which is/or has been, thus far, "inconceivable;" or "beyond our (developing) powers of comprehension." For example, what if at our current level of intellectual development, we cannot conceptualize other forms of cognition that exist beyond that which we already know. That is: what if our confidence (in our abilities) precludes us from pondering possibilities that humility itself would have and should have inspired us to ask? That is, namely: what if, scientifically we are just part of something larger than ourselves? And because we are "connected," to something larger than ourselves, is it possible that our consciousness and our intelligence is (in some transcendent way) in fact again; part of something larger than ourselves. And if we are connected to something else, does that mean that we therefore have obligations that (at this point) we are unwilling (or unable) to acknowledge. Or to put it differently, what if in trying to ascertain the origins of life, we are required to become compatible with a larger "collectivity" in order that we might be able to take the next evolutionary step for humanity's next (evolutionary) ascension. Moreover, what if evolution is a thing to be pursued, rather than something that's just allowed to occur? Would we have the wisdom and humility to even recognize such a worldly progression. And if this is so, are we then strong enough to comprehend that there may be a truth to the "existence of being;" that in the end, requires conscious participation? And what if the truth within "the science of life" should interfere with the societal way that things currently are? Again, are we strong enough to discover the purpose of life if it should mean that such a purpose itself places responsibilities upon our concept of "free will." So in a world that thrives off of injustice, separation, privilege, power and prestige; do we really want to know that science (or anything else for that matter) can show us another way? What if the science of evolution is not itself neutral on the way things ought to be? Does humanity (or any particular individual for that matter) have the right to be personally selfish or anti- evolutionary? And if so, what are the likely consequences?

And it should be recognized here, that to science these questions are not mere intellectual niceties. Rather it may surprise many of us that with regards to quantum physics such questions are the direct extended logical consequence of the core issues of the very existence of "matter" itself. For example, as already demonstrated there are things that exist without form, such as symmetry, harmony, logic, etc.; and then there is physical existence such as all forms of matter, from particles to all forms of wavelengths; and yet beyond these realities (that which has form and those things that exist without) we know that both these things exist, because these and other things have brought us to where we are. We exist; because (to crudely stipulate) inorganic matter evolved into organic matter, and organic matter gave birth to species diversification; and with the multitude of species, came the arrival of species consciousness (admittedly with varying degrees of consciousness amongst the species), so consciousness/ itself too, can now be said to exist as a reality without form. And that furthermore; through this thing we call evolution, other realities that we now know to actually exist, either i) always existed but in a drastically different form; or ii) these new things are truly new and that therefore their existence is proof that evolution has the power through (contact and motion) to create other things that are more than the sum of the original parts.

So the process of evolution itself might best be understood as the resultant consequence of matter and motion. And what proof do we have of this? Well the best proof is because, it has already happened. What we know is that things exist, and we know that things change; so evolution itself is real, and the creation of humanity itself, is proof of that. And the fact that: the process of evolution itself has performed this transition, through the powers of matter and motion is proof itself that these two things are part of the equation in discovering the origins of life. But the ultimate question (for now) within quantum physics, still remains: in the beginning, which came first; matter or motion? Because: the answer to this, might determine a still deeper understanding for the origins of life.

Eternity: A Concept of Indestructibility

So again, with respect to our analysis at hand as to attempting to ascertain the purpose of humanity, the answer between motion and matter and of the two which came first, may be of no immediate need for resolution (for now); because of the evolution that has already transpired, and because of the course and trajectory that our earthly existence has already put us upon. A trajectory of creating consciousness and all the attributes that have followed, such as deductive reasoning, logic and mathematics, which has provided for us a geometrical road map for what we might need to know to answer our more pressing concerns. But also; as an academic exercise however (for now), it may also be informative to recognize that our search for the origins of life ought to start at the subatomic level because: should we ever arrive at the truly smallest single particle to exist, a particle so small that it cannot be broken down in any further way at all, such a particle would then be theoretically, completely indestructible. And the consequent reality of conceptualizing such a indestructible solitude, such a solitude may then be demonstrative of a base reality. A reality so indestructible, that (short of its own destruction) it could be perceived to be "eternal." And within this reality therefore, there might be two consequent routes of existence; the first is that nothing is ever indestructible, and that should the smallest singular particle ever come to be destroyed, that to be destroyed (or further broken down) it would have to undergo an extreme transformation into something else, unleashing so much energy (or force) that a new harmony of existence (physical or otherwise) might be the actual result.

Or the second potential route of existence, is to postulate that if something is so small and singular that it can withstand all levels of destruction/even time (and therefore can be considered itself eternal), such a thing as this, may then have its own force of attraction, that for all that is within its immediate subatomic proximity, that such a force could be further theorized to be the theoretical origins for the reality of motion (and even be part of the theoretical origins of gravity) itself. So (ironically/and hypothetically), a theory of "absolute solitude" may have its own absolute consequences and reality. And though such a reality (can currently) only be hypothesized, such an extrapolation may/or could hold the answers to the earliest forms of motion; and as such a theory could then be further hypothesized: that with the origins of motion, all the other forces (through particle motion) become possible. Including any motion properties, that might actually account for, and propel the origins of life.

CHAPTER TWO: RECOGNIZING COGNITION, PAST AND PRESENT;

With an Expanded Definition.

So what then does this tell us about our analysis of life? Well indisputably; we know as an absolute fact (despite any hypothesis) that life on earth has experienced particle motion, cellular creation, cellular separation and replication, organic evolution, species diversification and the arrival of the homo sapiens, all of this and more, throughout the earth's many millennia. This we do know. But with respect to ascertaining a constructive understanding of life and its potential meaning and purpose, one of the most pressing questions still remains (and springs from all that has been discussed thus far) and that is: is "life" itself a miraculous thing; or is it miraculously, a series of many different miraculous things? Or to put it another way, even as we conclude that earth's many varied forms of life can be traced back to a common heritage of origin with respect to location (i.e. the ocean's) can it be determined that the variety of life on earth should be analyzed all together (thereby recognizing some form of unity); or does it follow that our planet's "life" analysis can/or should be: species specific? That is; since it is we humans who are conducting this analysis, and since humans have been credited (by our own accreditation) as being the most advanced and evolved intellectual species on this planet, does this entitle us to separate the purpose and meaning of man as being something, if not different in kind; then at least something (again, dare it be said) intellectually special, from all other forms of life (again as we chose to recognize them).

As a Species: Are We One of Many, Or Are We So Special, That We Are Different?

So if we are to tackle the question of whether life is to be examined as a whole or in part; maybe again it is best at this juncture (if we are to have any humility at all) to look back and confirm once more what it is that we actually do know about life; as it has evolved and advanced thus far here on earth (again from our middle evolutionary perspective within time and space). Why? Because apart from the fact that "we humans" are only one of many species on this planet, we have to remind ourselves that we too are intricately intertwined in the evolutionary equation (that is both ongoing) and evolving at the same time. It's like trying to assess the significance of a particular team sport, as an active player, on the field; while the game is in progress and the (invisible) referee, is changing (or evolving) the rules while the play is underway. And that in a nutshell, is our problem. And, oh yes: this evolutionary game of which we speak … it is most definitely, a contact sport.

So since we are but one species among many, are we entitled (by this fact alone) to contrast ourselves from the other species as we perceive them; and if so on what basis is this separation justified. And even if that separation can be substantiated, conveniently, such a separation creates for us (suspiciously) two self serving choices for us to consider. First we can perceive with our acuity and accomplishments that we are the most advanced of the species and that therefore there is some sort of hierarchy to be realized (and therefore all other life forms / both within humanity and otherwise, are open to be exploited); or alternatively, as "one of many species" on this planet (advanced or not); we can interpret ourselves as distinguishable from the others with our various capabilities, but that "through a perception of global collectivity" (again advanced or not); such capabilities through a perception of interconnectedness; also presents (for our species) an array of unique possibilities, benefits, and privileges. Privileges, that in turn, entitles us to assume some sort of paternal/custodial status that in turn allows us to presume that we might know what is best for all the other species. And again, either way; both perspectives are conveniently (for us) self aggrandizing, if not, self serving.

So let's be clear; a decision on this very issue, has already been made. We have decided that by choosing to engage in the analysis that we are currently discussing, meaning: a search for the purpose of life, that we as the analysts; we have assumed that we are (in fact) up to the job. And because we have ordained ourselves as being up to the task at hand; that therefore we are distinct and special enough, that we are therefore worthy of separating our human purpose throughout this enquiry, from other possible holistic considerations. Now both options namely that: i) we are qualified enough to ask the questions we are asking; and ii) that in conducting this enquiry we are entitled to separate ourselves from all the other species, immediately presents to us (as analysts) a circular form of foundational reasoning that separates us from the other species, because of our supposed advanced form of cerebral intellect. And, to give us our due, our circular argument does at least create a self fulfilling conclusion, namely: the fact that because humanity has unique cerebral capabilities these capabilities should justify why we should be separated from the other species. And conversely, by emphasizing the need to separate ourselves from the other species, is (supposedly) the best way to prove our unique and superior cerebral capabilities from the other species. So our very presumptive circle becomes presumptively complete. So whatever else might be said about humanity, and our achievements thus far; undeniably it is our intellect (our cognitive evolution), that does appear (through isolation) to be the thing that most has distinguished us from the rest of the other forms of life here on earth. And so with that supposed fact at hand, it makes sense that we explore the concept of cognition and the development of cognition (and in so doing we should be willing to expand the possibilities for cognition) beyond our current understanding of cognition within the organic.

And this means that not only do we need to (again) continue to perfect our search for knowledge and our understanding of all things; but in contrast we should come to the realization that as part of a evolutionary process of which we are a part, that through our potentially advanced cognitive abilities, that we humans might also be able to make a contribution (through consciousness) to the evolutionary process itself. So since we are actually part of the subject matter being analyzed, the very point of evolution might be that through our cerebral abilities, such intellect could potentially (as we have tried to do) give language to the articulation of progress for life itself, So, much like math provides description for symmetry, and by using such things as musical harmony to show us that harmony itself has always had a its own separate existence (possibly) in all things; the human ability to conceptualize these (and other empyreal) things; that by giving language to the empyreal so that it might materialize conceptually; and in doing so, we might also be giving the empyreal the ability (at our level) to "harmonize" back. For instance in the earlier parable of a team sport analogy taking place contemporaneously on our earthly "evolutionary field," in which we (in time and space arrive somewhere in the middle of the match); what if, we were to hypothesize that we might, through our conduct, demonstrate our conscious acknowledgement (to anything and everything) that we are indeed conscious of the fact that we are only part of a larger whole; and that in the presence of invisible (far more powerful) force/or forces, that we (despite our cosmic/youth) might be just sophisticated enough now to let those forces sense, that we understand that other forces (and life forces) beyond ourselves, do exist.

And this is important why? Because if we apply the axiom that the simple prospect that by placing a subject under observation, that this fact alone is enough to alter that subject's behavior. And although it is unclear how far this axiom may go in its applicability to all things, inert or otherwise (on this the science is unclear); we do know however that humans behave differently when being observed, and on this what might be more interesting is (not the question of whether we change our behaviour when being observed) but whether the "forces of existence" themselves behave differently in response to our changed behaviour because (we simply believe/ correctly or not) that we are under observation. That is: can the reverse be possible that because we think we are being observed (and thereby we materially change our behaviour), can that measurable change initiate otherwise dormant or inactive life to itself respond in any measurable way. That is, once we obtained consciousness, was it possible that our conscious acts (without deliberately trying to do so) were able to activate inert life properties in other things. For instance, a logical comparison might be to consider the deliberate use of consciousness when we use chest compressions (as an outside third party) on an inactive heart, and then through cardiac memory, a restart to a dormant heart takes place through the physical act of applied external transferred kinetic energy. So in such a case where we know that a dormant heart can be stimulated back to a self sustaining life support system (provided however, that the requisite energy is applied before there is too much degradation through the passage of time) has occurred. Is it possible (at the subatomic level) that our actions to behave differently (because we wish to discover a harmony of any sort with our subatomic "strong forces,"); is it possible (deliberately or not) that we could stimulate such forces (within the nucleus of our atoms) through our actions; where time (as we perceive it) at the subatomic level may have little or no degrading consequence.

So through the logic as expressed above can it be postulated that through the process of changed behaviour (that results in material change due to kinetic energy/or other things) can such a change in turn stimulate unknown properties within the atomic structure of what was thought otherwise to be an entirely inert or dormant particles. Or to put it differently: can motion, collision or the exchange of kinetic energy (or other things), bring unknown nuclei properties at the subatomic level, to become activated, and if so could those properties have an (altering or contributing) effect on "life?"

And there's the rub, if it is possible that there are life contributing forces that exist at the subatomic level;

then is it not also possible that this realization itself, is an evolutionary step in the intellectual capabilities of mankind, because the form of life that is now therefore ultimately being conceptualized, is that there may be life propelling properties within seemingly sedentary and impregnable particles of matter, that without (or in the absence of external) contact, motion or (some sort of) stimulation; there would be (within that matter) supposedly no life at all. And such a realization as this could change mankind intellectually as we would now understand life from a new perspective that more intimately connects us with all things regardless of their form, or their currently perceived life status. And this is of course is a quantum leap (and yes …the pun is intended) from the earlier organic comparison over the debate of questioning whether there are inert or dormant life held properties in other things, such as viruses. In this subatomic analysis there is no organic DNA code to be altered; in this case scenario the reality of DNA comes much later. In asking the question however, could the purpose and meaning of life stem back beyond the transcendence of the subatomic, allowing us to ponder the possibilities that the origins of life might be found anywhere, everywhere, and within anything; including that matter which is perceived to be sedentary, impervious or "lifeless."

And this intellectual proposition, is itself; evolutionary. It is the conception that there may be life capabilities not only in the things that appear sedentary, but also raises the question of whether there could be formulaic properties for life's construction in things that exist without form, such as symmetry. And this raises two fascinating possibilities. The first is that since we know things such as symmetry, precision and harmony do exist without form, and therefore (from our human perspective are therefore conceptually eternal) what reconciliation is there to be had conceptually; when the eternal is to be rationalized with subatomic forces? Is it possible that the one could have a formulaic (design) influence on the other? Or, is it more likely to suggest that such eternal things came from within the subatomic realities of either the "strong forces" themselves, or something beyond? Either way; a empyreal design equation arises of whether at the subatomic level (or still something deeper) do the eternal realities (without form), arise from the forces within those same subatomic realities; or could the eternal realities without form have some sort of external design influence over the (subatomic) strong forces as they interact with other particles and other realities? And again, either way, could such a rationalization, from (which ever school of thought one might choose) help us in our analysis for understanding the origins and purpose of life.

And yet with these questions about the subatomic influences possibly exerting themselves on life; we find the scientific analysis tends to disregard and refute the possibility that those forces are part of us. That is: by approaching our analysis through (seemingly sterile) clinical observation, we thereby separate ourselves from the analysis at hand, we objectify the search as if it were something to be observed as opposed to something that is to be experienced. And this objectification becomes problematic because it reaffirms a predetermined bias that those things that are empyreal, are to be limited to being seen merely as that which exists without form, and nothing more. But by prejudicing our analysis to that which can be empirically concluded through clinical observations, electron microscopes and proven mathematics, we presume to know the limits of the very things that are being questioned. We flatter ourselves with the assumption that the choice to make such a study is itself a definitive conscious choice (made by us solely through our superior consciousness of thought); made completely devoid of any empyreal influences from other realities. In this conclusion we simply presume that through conscious choice, that such decisions made, are made because intellectually we are uniquely (and solely) "capable." We reject the notion that although we realize that other empyreal realities do exist, we reject the notion that such forces may have had an ongoing influence on the decisions which we believe we independently made. On this our science community has convinced most of us unnecessarily, that: because we perceive ourselves as the most intellectually capable and advanced of all the species, that therefore, in the study of all things ... we humans are intellectually: "all alone." Excluded in our analysis is the possibility that our intellect and thereby the decisions we ultimately make are in fact the extension of the forces that brought about our existence, our consciousness, and our consciousness of thought, to begin with. Where conversely it is completely plausible that our intellect (and by extension consequently our thoughts and decisions) themselves may have their initial transcendent origins inspired by the energies/ forces and motions that take place at the subatomic level.

This premise however, the premise that we are the only intellectual species capable of contributing in an analysis of the meaning and purpose of life, should be immediately recognized as being unnecessarily self imposed and needlessly over restrictive, even if it is true that as a matter of intellectual dialogue, there is no other species input except that which can be clinically identified by us (as observers) as each species goes about their daily activities (like an ant hill in a forest/ or an ant farm in a clinic); supposedly and apparently oblivious to the fact that by virtue of doing so, they are actually "visibly" communicating and supplying (to any interested observer) the visible rational for their own unique behaviour. Put simply our own search for intelligent life may be blind on two fronts; first, we refuse to see what is immediately in front of us; and second, we even refuse to contemplate that there might be a greater intelligence right in front of us, here on earth; which exists as an:"intelligence in the aggregate." An intelligence, that: may have been at play as an invisible precursor, a manifestation of subatomic properties which influenced (or at some incomprehensible level/ even "inspired") in advance our actual intellectual decision to conduct this enquiry about the meaning of life in the first place.

Moreover, when a species of this world goes about their daily activities, for all to see: that it just may be that: that particular species (in doing what they do), are in fact, doing their part to take "everything" to the next level. And this does not necessarily mean that they are following a preordained plan; it only means that in doing what they're doing; they are doing their part, for a greater whole with what evolution has thus far, enabled them to do. And at this evolutionary time; their part, may simply be to try to survive and evolve further, through a proven process of trial and error. And, not surprisingly, this process of trial and error; is exactly what we believe the process of evolution itself is trying to do: it is the process (of trial and error) to overcome any and all of life's immediate obstacles (physical or otherwise) so that life itself can survive and potentially improve. So as a matter of specifics; the bees in a hive do what they do, because they are part of a larger whole; and it is completely unknown whether they have any comprehension of what role they play or the service they perform. But we must be careful in deciding what we perceive them to know; because what they comprehend might be entirely different than what we understand. Moreover the indifference of many collective species display, when confronted by our (human) existence, too; may be equally catastrophically misunderstood by us, for such species may again on any number of (wavelength or empyreal) levels; may be just demonstrating to us their own form of colloquial dismissive(ness), as they wait impatiently for us to realize our own part; maybe in the hope that we will eventually "catch up," and "get with a (larger) program." ...So, such might be the folly of our superior, human intelligence.

But again, the assumption that intellectually, we are: "on our own;" is senselessly one dimensional and an affront to what is actually taking place directly in front of us. Varied species, in doing what they do, conducting their daily activities in front of us, as they do, and for all those to see; is itself (whether they mean it or not) an actual process of communicating with us (and with potentially "anything else?") that chooses to see. Also, we cannot rule out that they too (other species) are not also (at some level) somehow cognitively aware that their activities too, are either beyond themselves important, or that in fact aware that they are being observed, as being part of a larger collective, a collective that is intuitively aware that they are part of a greater whole.

Knowledge: What Is It?

So what do we know? Well we know that (as a matter of process) if mankind is to have the most comprehensive understanding of life, such an understanding will require both that our pursuit of that understanding must be pure; and that if what we discover is true, than such discoveries ought to lead to an infinite transcendence of further discoveries where the most recent discoveries will vindicate the truth of the discoveries made before. And such discoveries will then lead to more knowledge and from further knowledge, then maybe we too can contribute to both our own evolution and maybe even assist (through connectivity) in the "conscious" evolution of all other living things.

So if this is to be our pursuit, then we have to come to the realization that if it is within us to "intellectually assist" in the advancement of evolution itself, then in order to do so (for the purity of our results) we need to first purge ourselves of our inherent biases and shortcomings; and once that is done, only then can we recognize our own potential to make a contribution to the perfection of knowledge itself. So how do we do this? Well to do this, we have to do both these things simultaneously by giving knowledge a (materializing) language, that has within it both sufficient expanse and precision to ensure that the linguistic representations are accurate, complete, thorough, and as uncorrupted as possible; so as to not become potentially misleading. Why? Because it is reasonable to assert: that the most accurate perspective of life must also be (by definition), the most honest and unselfish version for what life might be, because if we presume any shortcomings or limitations, then our analysis and search for the true forces that propel life might well be flawed, because our underlying presumptions were flawed to start with. So the paradox begins: if we are to perfect our understanding of life we must first be prepared to improve upon ourselves, if we are to achieve a more perfect understanding of "life". Or to put it differently, if humanity wishes to have a truer understanding for the meaning of life, we as a species must necessarily evolve to escape our more primitive and parochial past, a past that saw our intellect diverted by economic and geopolitical conflict, rather than realize the potential for quantum intellectual harmony.

So whether individually (species by species), or collectively, as mankind; humanity's pursuit for a consistent purpose of, or/for the meaning of life (in its broadest sense); lies with our perfection of the knowledge that accurately explains the basis and purpose of evolution itself. A knowledge that becomes infinitely self fulfilling as greater knowledge allows humanity to intellectually evolve thereby sustaining still further evolutionary comprehension which in turn should activate an infinite ascending spiral of transcendence as each intellectual evolutionary step serves as the cerebral foundation for still further transcendent advances in intellectual comprehension... and so on. Therefore the knowledge and the answers that we are theoretically looking for, is that which ultimately forecasts some degree of contribution to the overall evolutionary process itself. In short, in order to take the next evolutionary intellectual step, namely that there is in our evolutionary process, a propulsion towards an overall collective recognition; we will have to evolve from our current preoccupation with, "self and having;" to one of: "us and giving."

So If we are to succeed in simultaneously purifying our scientific method while contributing to our own (hopefully) unselfish and uncompromised cerebral evolution; such assistance in either case will be had through our conscious contribution to life through "knowledge;" presuming of course, that it is within us that such "knowledge" can be translated into a helpful language. And, it makes sense that if humans are the most intellectual of species, then it does seem to befall us to perfect this language, but as the translator and interpreter of this language it will also be necessary to perfect our understanding of other intellectual realities such as reason, rationality and consequence, and of course with consequence, we will have to deal with (eventually) the reality of our own mortality. And who knows what a deepening perception of life will result from a greater understanding of that?

But regardless of the above, always present will be the question: why us, or why humanity? Well, by virtue of our shear existence (and because again, we arrived in the middle of everything); humanity has the ability (as first discussed in this analysis) of having the benefit of being simultaneously intellectually capable of looking at life both from an individual perspective, while simultaneously having the benefit of looking at life from an ever broadening collective analysis. Collective, both from the human perspective (beyond the individual), and collective; from an all inclusive species perspective. And this multi dimensional perspective grants us the unique ability to measure life's evolution both on an individual basis (within each species), while also watching for whatever collective contributions we humans as a species might make And then there is the contribution of what we in collaboration with all the other different species might make to the overall evolution of life here on earth. So it may not be a question of why us, why humanity? It may be that homo sapiens were just the most immediate and efficient life form so far, to come to this conscious awakening, and that we have yet to prove ourselves as being the most promising mechanism to give life to an empyreal consciousness; because again, in almost every conceivable way, ... we are at best, still in the "middle" of everything.

But if a case is to be made for ourselves; that we are indeed a most appropriate species to carry the mantle of a perfected consciousness; it may well be our unique human ability to see contributions to life; both from an analysis of individual species (in their particular advancements), to the larger more whole sale advancements made by life collectively to the evolution of life generally. And this might be one of our greatest strengths. So even if there is a contribution in the smallest of measures from an individual within a species, or from a species generally; such a contribution at a particular time and place (even by accident) can make a cascading contribution to not only an evolutionary process already underway, but also grant an ever greater understanding for the overall evolution/or transformation, of all things. Put simply, our greatest contribution may be in giving a conscious appreciation of a ripple effect in evolution (even through the proven ripple effect that occurs through natural selection) may have, or could have; come from anywhere.

Therefore what we might surmise about life, is that (by definition) when something endeavors to evolve, whether it be at the molecular level or as part of a larger organic system; then from this, can it not be said that such a thing (for a lack of a better word) is at least: "transforming." And of course this line of logic then becomes highly significant because maybe all things that are transformative may have the possibility to contribute to life even though that contribution might be difficult for us humans to perceive.

Therefore with this simple (possibility) assertion, we can still advance our analysis to the next logical enquiry, even in those instances where the last enquiry in logic, just left us with a potential "possibility." And that next logical enquiry (in this analysis) is that when something can be determined to be (evolving and/or transforming) can it not also be surmised (or at the very least not ruled out) that:

that very same thing, may also be said to have some sort of base: "cognition," or a "cognitive property." A cognition which, may only stand for the fact that: even a sedentary thing, may have the working ability in physics to repel its own obliteration, termination or expiration. Such properties within something solid, could then be (defined) to have at least affirmative "operational property/ or properties" to resist becoming broken down indefinitely to the point where it might be said to have no further physical properties at all. Now typically such an invincible existence is presumed to occur at the atomic or subatomic level but no matter what level of existence physical matter descends to before it starts pushing back, refusing to be broken down any further, is not then logically such a point of resistance (a point of "self preservation,"); and is not such a form of resistance, not itself (some sort of power source); or some form of propulsion where the origins of life, might begin? It being contended here, that the origins of life might best begin where destruction ends. Such an origin of power could be seen to exist where we might find the internal bonds which keeps any base integrity intact. A base power source/ force upon which other bonds can build a foundation. And if this is so, then does it not befall us when analyzing sedentary/or solid things (using our definitions and language) for us in choosing our language to possibly describe such phenomenon to embrace that terminology that might best provide for such/ and all possibilities?

In searching therefore, for the origins, meaning and purpose of life; should not our language, our choice of terminology be such that our words anticipate and embrace any and all possibilities? Such is the case here; where the two words: "property" and "cognition" could be used potentially interchangeably; but by themselves this is currently unlikely because, one word "property" doesn't do enough; while (typically) the other word "cognition" does too much. Consequently, when discussing the unknowns of atomic structure, we have to understand that some words carry with them needless connotations, and other words do not. In this case, the term "property" carries with it little or no perceived connotation towards the self maintenance of any "thing;" and the word "cognition" carries with it an automatic dogmatic interpretive (reactionary) connection to the word, "consciousness." But such a spontaneous reflex need not take place if rationally one is to understand that when dealing with solid sedentary atomic structures the word "cognitive," can and should, simply infer the resilient forces that simply allow a thing to continue to be. And the benefit of having such a new working definition (for a base cognition) is that these abilities could then be embraced as a positive characterization of unknown potential where such atomic matter might be perceived as a precursor to be later aligned with organic cognitive positive activity. Such flexibility in language could then more accurately capture a transitional process where the concept of maintenance and self preservation would not be seen as alien concepts to one another. Such conversational use of these terms interchangeably could then embrace the essence of a transcendent atomic activity that is itself, (at least) akin to (dare it be said) some sort of self preservation, or even a physical solid formulaic foundation (or code) with information capable of being transferred for the "achievement" of future evolution. And notice here that the choice of the word "achievement" in relation to future evolution, is a discernible choice in contrast to the use of the word "purpose" in relation to future evolution, because the word "achievement" comprehends the existence of a cognition for our planet's evolution as a collective that already exists in the "aggregate," where the use of the word "purpose," might assume too much. It is the difference between a life force cognition aspiring towards some sort of consciousness (as our earth has already achieved); as opposed to a consciousness of any particular sort, already being predetermined.

Consequently, if we broaden our definition for a greater understanding of life to include that which assists in the transformation (or the preservation) of things; such a base phenomenon could be viewed more positively and consequently work its way into being more worthy of our observation. Or to put it differently, can we (continue) to assume scientifically that there's no "cognition," when we actually know, through transformation, or self preservation: ... that there's (at least) therefore, possibly; ... something. And why do we know that there is indeed "something?" Because millions of conscious species currently, and in the past, have already existed; and because we know at some point, and for some things: physics does indeed, (and has already) transformed (over millions of times); ... into chemistry

Cognition and Consciousness

Or to put it another way, we would normally recognize the perception that cognition and consciousness might be the same thing but representative of two different extremes; where one is perceived as the precursor to the other, but such an understanding need not be the case when the concept of "cognition" is used to describe atomic and subatomic activities in a sedentary body that naturally experiences an entirely different reality. This is to say, that where cognition develops in a favorable and inclusive environment, such an environment might allow for the evolution of cognition and then later the further evolution of enlightened consciousness (for a particular species) when the forces for such evolution are conducive to that sort of intellectual development. But where those same initial (atomic) cognitive forces (coming from the exact same atomic sources) are exposed to/ or experience a vastly different, opposite and hostile environment; than those same atomic /or subatomic abilities, may have the abilities/properties/or cognition to only respond to that environment in a way that is only compatible with that hostile environment. But just because one form of development is more rapid (in activity) and seemingly more advanced because it is nurtured by a conducive environment (as opposed to the seemingly more limited development which might occur in a hostile environment); this does not mean that the original (cognitive) origins that came from the same atomic matter are not still derived from the same source of initial "cognitive properties" that existed within that same atomic matter. Consequently, two completely divergent paths of existence can emerge from the same originating "cognitive" properties.

So in such a contrasting situation as this, what might be most helpful to our analysis for the purpose and origins of life is to realize that just because there are completely divergent outcomes from the same primary elements, when the (predominantly) same atomic (and subatomic) matter is/are exposed to two starkly different environments, this does not mean that the same initial life creating properties were not at some level present (all along) within that same atomic element. And the fact that atomic and subatomic realities are capable of producing such divergent outcomes of existence with contrasting varying degrees of both property cognition and the much later (potential) for consciousness, does not mean that the sub atomic characteristics/or properties were not at the root of all transformations that eventually occurred in two entirely different environments. Thus by expanding our definition of "cognition" to simply be open to the base properties that allow for two diametrically opposed outcomes of transformative development, from the same atomic element; by adopting an expanded definition for the potential of "cognition," we broaden the concept from an intellectual thing (what we would typically contemplate as a cerebral thing) to also include the foundational properties of quantum particle theory.

And this recognition may demonstrate that we may have an opportunity to rethink transformative things; where such things may have some form of "operational capabilities" simply because they (through time, motion and collision), such things have the ability to actually change. Now this does not mean that something that is transformative is capable of actually transforming or evolving itself. But this fact doesn't mean that it didn't; or that it can't contribute to the life of other things. It only means that as a matter of definition and recognition; that the concept of evolution as it stands now, may need to be open to include atomic and subatomic contributions as precursors for what later is credited to be "evolution" itself. And it may be that these (perspective) attributes or properties, may themselves prove to be the most revealing; with possible insights and revelations into the origins of: "life" itself.

An Example: Upon Reflection

For instance, it is almost common knowledge that diamonds are the result of carbon, being exposed to extreme heat, pressure and time. And that through this transformative process, these three external influences, have sufficient force to compress carbon atoms to the point where their proximity to one another, has the transformative ability to create a crystalline reality. A mineral so atomically condensed that it becomes one of, if not the hardest mineral; known to man. And yet not only is this mineral one of the strongest minerals that is capable of cutting through all other minerals; but in this condensed form (as a crystal) this atomic structure becomes completely translucent so as to have the additional capability to conduct and dispense light; showcasing the array of differentiated wavelengths and colors within the white light spectrum. Carbon density therefore, under these extreme influences, creates this contradictory reality of both density and transparency, within this particular transformation. A transformation, from an initial stasis, to a super mineral, with light altering capabilities; and a sustainability of almost infinite longevity. Such are the capabilities of carbon when it is pushed to the external extremes that would annihilate most other matter; however with carbon there is the creation of an alternate sedentary existence and yet through this violent transformation, where carbon (exposed to these conditions) successfully transforms itself into another reality, a crystalline hardened reality where (with its new capabilities) it can process and interpret light conceivably, for eternity.

However, carbon (also under different conditions and stimulus, in contrast to the above) is also the foundational element for all biological life, where (in combination with the likes of hydrogen) carbon becomes a foundational element for the existence of such life as the "Immortal" jellyfish. So how can this be? How is it that the same foundational element can reconcile these two almost opposite extremes, unless it is understood that at the atomic level or the subatomic level; elements have capabilities of, such transformative properties that two such divergent realities can be harbored by creation, from the same atomic source. The properties capable for providing for two such divergent outcomes commands that within itself the element carbon has the ability to exist in a translucent hardened sedentary state which also has the abilities of conduction and refraction of light while existing indefinitely in that crystalline state. But quite astonishingly, this same element can equally adapt to being a base element sponsoring basic biological functions that also appear to aspire similarly in favour of an indefinite mortality within the immortal jellyfish.

So logically, what we are left with is two different forms of existence, which harbor a vast array of very different capabilities and attributes, but yet stem from the (instructional) properties of primarily of much the same atomic structure. And even if we credit any other elements involved in these two separate (diametrically opposed) outcomes of diamonds /and jellyfish, even with the influence of any other elements involved; we are still faced with the fact that in both the sedentary and the biological form of existence, there is the same common denominator of the primary element of carbon. And this carbon element is therefore the basis of two entirely distinguishable forms of existence and this means that carbon's atomic ability to adopt an impregnable existence (when the hostile outside forces are there); or when induced into a biological existence; can also form an almost translucent cellular membrane as a biological animal (a hydrozoan turritopsis dohrinii) that has the capability to self regenerate in such a way as to provide for its own long term (potentially immortal) continued existence; surviving in perpetuity in a cellular form. Thus the properties or the forces that enable one form to result in a sedentary (mineral) state while acquiring and maintaining light dispersing capabilities indefinitely; such an element also holds the properties in contrast to allow for another form of existence to become alive with still further evolutionary capabilities to possibly follow. Put simply, the element of carbon is proof that some elements can become "rock" or "life," but more surprising is the fact that in both instances such an element also has the ability to present either version as being timeless, to the point where both can appear to have the ability to be eternal.

So once we have expanded the definition of sustaining properties to include the sustaining forces of that which may be sedentary and impregnable;

such a definition will allow us to understand that such alternate transformations may mean that at the subatomic level that the power to self sustain may have forces that, under different environments; release different capabilities. For instance where the external forces of heat, pressure and prolonged periods of time may force subatomic results of condensed atomic existence that becomes binding; the alternative may be true that in a conducive environment of a lack of pressure and a nurturing environment where there is an abundance of space and other elements; under these more hospitable conditions, the subatomic properties may adapt/or transform in a more expansive way (beyond self sustainment) than that which was forced upon it under other crushing circumstances. So, just as time, pressure and heat will have its consequences, so too may atomic exposure to facilitating atmospheres with varying temperatures, may have theirs.

Is Intelligent Consciousness, Really More Intelligent; Or, Just More Consciously Active?

But if all self transforming things can potentially contribute to life , than what does that tell us about our arrogant presupposition that "intelligent life forms" are themselves advanced life forms because they exhibit "thoughtful conscious intelligence," more advanced then supposedly those cognitive life forms that appear only to be cognitive for the purpose to survive. It may be that what we should recognize is that "intelligent conscious cognition" is not necessarily a superior form of cognition, but rather only an evolutionary intermediate (intellectual/building block) stage of a larger (all encompassing) cognitive life continuum; an understanding of life which includes a concept that a form of cognition may exist to elevate one level/or form of cognition to the next. That cognition and consciousness should be conceptualized less horizontally and more vertically as a process of ascension. That a previously developed (base cognition) within any given thing, is a foundational base that has the functional ability of connecting a base form of cognition (previously achieved / like a solid foundation to a house) with any perceived higher consciousness (like the raised scaffolding of the next floor) within any given species. Put simply, that a species pre-existing cell structure can pose as the molecular structure that can nurture the creation of further neurological/ brain activity. The premise being put forward here: is that: intelligent consciousness is not necessarily a more advanced form of cognition but rather more an internal transitional stage of an evolving (ascending) cognition; as one form of cellular construct can pose as a base towards the next generation of (possibly higher DNA) advancements.

A crude conceptual representation of this paradigm might be to view the human ascension of cognition as one might look upon an x-ray profile (a side view if you will) of an ascending blue print of a high rise building. For instance, if one were to view one level of cognition as say the lower (concrete) floor (a floor of some significant depth and thickness/ representing the many accomplishments of evolution to date); and that above this floor is an open space where the day to day neurological activity (of such things as the intellectual creation of theorems, art, music, mathematics and calculations, etc.) takes place; and that above that, is another seemingly sedentary concrete ceiling; then in the ascension of cognition from one floor level to the next, if the internal DNA transition is seen as a spiral (intellectual helix) staircase leading from one level of supportive cognition to the next; than it becomes easier to comprehend that one lower level of cognition may appear as being more dense, less active (in comparison to the hectic cerebral activity that is taking place) in the open space between the two floors. But just because the appearance of greater activity through the frantic firing of synapses in one's brain may make it appear more functioning, and therefore more determinative than the seemingly less active, more dense, cellular cognitive creations already achieved in the flooring below; such hectic neurological (open space) activity should not then be presumed to be more (cognitively) advanced than the more dense, less visible, (concrete) cellular cognitive activity that has already (chronologically) occurred. To put it simply: we humans, through our intellect, may be trying to consciously understand through mathematics and quantum physics all that is going on (and through our consciousness we may have actually been able to have made a contribution) but such achievements should not lose sight of how infinitely immense and indeterminable (not to mention multi-dimensionally) vast, the rock solid achievements of the universe have already been.

An example of such a deceptive duality as this might be demonstrated in nature, at the cellular and metabolic level involving the evolving DNA adjustments made within the species of reptilian alligators. For instance North American alligators have the evolutionary ability to sense imminent changes in temperature that may lead to the flash freezing of their habitat water beds. So once the alligator senses that there will be a drastic drop in temperature to possibly cause the alligator to become trapped beneath the ice, the alligator will have the cerebral knowledge to push its breathing nostrils above the water line so that once frozen the alligator will have access to air and oxygen though the rest of its body will remain submerged (and even somewhat encased) beneath the ice for a length of period of time potentially longer than a week or more. And although the knowledge and choice to secure access to breathable air can be attributed to the cerebral activity of the neurological brain, the alligator's metabolic ability to have first: the tactile ability to forecast severe weather changes (in advance); and then, second: to thereby reduce its heart rate to three beats a minute in order to accommodate an extreme version of urgent hibernation (in contrast to that of say: bears); such sophisticated adaptation can only be attributed to the evolutionary genetic DNA information developed over millennia where the reptilian genome adaptation has successfully brought about the equally compatible cellular metabolic mutations that did in fact allow such alligators to survive.

The point being made here with the above example is that: the conscious awareness to maintain one's nostrils above the ice, while such awareness is itself a deliberate demonstration of consciousness (if not, even an actual intellectual choice); such cognitive cerebral abilities seem however to pale in comparison to the storage, and the metabolic adaptation of cellular information (which through harsh mutation) successfully redesigned the alligator's anatomic abilities to provide for the immediate survival (and therefore the ancillary chance for future species survival) when faced with an imminent threat of death or extinction. So maybe we have to simply acknowledge in our most rudimentary understanding of life; is that all evolving things have some level of cognition, and that cognition itself could possibly start at the cellular level (or even the molecular level), and that what we currently understand as cerebral intelligence may be nothing more than a rather limited functional construct, a staircase (if you will); built on top of a far more sophisticated (but seemingly less active) consolidation of physical (cellular) informational matter. The cellular matter that allowed for the adaptation of the alligators cardiovascular and respiratory mutations to include the ability of extreme hibernation in order to survive, occurring at the cellular level, independent of the (brain's) cerebral intellectual function that was just smart enough to poke its nose above the waterline. Therefore cellular matter which (though seemingly sedentary/in this case almost frozen at any one given moment); has therefore through DNA evolution; not only created "evolutionary life," but may have also radically increased the very rate of transition into an exponential expansion of (changing) evolutionary "life." So what is being suggested here is that the evolutionary development of transformational DNA, may have been (in the past) to the cause of early evolutionary development, what cerebral intellectual (human) contributions are to us today. Or to put it as a cartoon caricature, before there were scientists with lab coats and test tubes, there was the cognitive (intellect) of a reptilian DNA scientist holding a helix (or two), looking for a solution.

So where does the above logic leave us. in our inquiry as to the existence of life and our understanding of not only the potential origins of life, but also our base definition of life. This is to say that by trying to understand life, have we not now stumbled upon the notion and the (possible) logical conclusion that life (meaning all life) may exist as part of a deeper continuum, than just (say for the moment) a compartmentalized intellectual belief (in understanding life) as existing primarily in some form of particular species organic matter. Life, from this perspective, may best be perceived as an ever lengthening or deepening progressive continuum projecting in (at least) two directions; one, towards the infinite transcendence towards an analysis of the subatomic character and properties that exist within matter itself (much like negative integers are to the opposite progression within the concept of math); and two, the opposite direction towards understanding the potential for life within the infinite and farthest perceived realities within our universe. And although this description (or intellectual construct) may be easily criticized as being unnecessarily simplistic and one dimensional; from a foundational perspective, starting from a linear (singular) progression from the subatomic existence of all matter to the furthest reaches of our galaxy, such a rudimentary (albeit linear) understanding may allow us (at least for the moment) an initial "step in a staircase" upon which we might later achieve a deeper and more dense appreciation for all other multidimensional cognitive possibilities.

The Transition from the Perceived Impregnable to Biochemistry

So if the origins of life are to be understood to be that which is capable of even the most basic achievement of transition, that its earliest state may be nothing more than some sort of movement (internal or otherwise) then the obvious question then becomes where does evolution (or life creating adaptation) come from? Well the structural answer to this is that: yes the structural origin of molecular evolution does seem to stem back to those instances where atoms realign with each other to form new or different molecular compounds, but it also possibly takes place (depending upon one's definition) at the atomic and subatomic level where supposedly stable atomic matter becomes activated or even altered by either external (environmental) interaction or stimuli. One such example of such interaction occurs when electronegativity accounts for those atoms that can draw electrons from other atoms and this ability proves that properties without form (that being the electronegativity, itself) not only exist but that they do in fact physically and tangibly matter. But it is basic science that an atom's nuclear structure (though typically stable) has a corresponding equilibrium of corresponding particles of electrons to protons; and although it may be that we may never understand why individual atoms have their specific number of electrons (within each element), the fact that we know this to be the case and that there are forces such as electronegativity to disturb this equilibrium, is itself proof that there are forces and realities (without form) within atomic theory that transcend physical realities that just simply cause/ or force particles to move.

So proof, that such intangible things are real; warrants computation into our ever expanding analysis for what we should consider possible for their contributions to life. Why, because what is being suggested here is that things (without form) such as electronegativity and other things such as the unknowns within an electron's cloud; may (or may not) have unknown properties that influence the actions and reactions of a particular atom/or other atoms, in ways that have as yet, not even been conceived of. And the fact that this, a thing without form is itself active and therefore potentially influential; this forces us to recognize that we simply do not know what actually is the driving force behind the existence of life. It may well be, that the unknown origins of life; may extend back to the cause and effect of motion and collision between the impregnable and that which moves at the speed of light. Indeed the question may be symbolically represented by the equation of: "what does happen when a (supposedly) impregnable object meets an irresistible force;" at either the atomic or subatomic level.

So (for now) currently, with the facts that we do have; "life" here on earth is attributable to the physical interaction of matter; that from the realities of atomic interaction, movement and connection (beyond the atomic) … where molecules are formed. And then, through molecular interaction, more complex organisms mutate and continue to form, and from there through millennia and an indeterminate multitude of further complex combinations, a continuous evolution of varying species, life (in its most visible form) has arrived, and this has become thus far, our successful reality. And the fact that, as humans; we flatter ourselves that through our intellect that possibly we ourselves might have something to offer;…"well isn't that (just something) special."(D.C.)

CHAPTER THREE
LIGHT: AND HOW IT HAS BEEN ...

So in the beginning, we started with the question what is the purpose of life, and from there we realized that in order to discuss the purpose of life we had to have a better foundational understanding of what life actually is. And through the analysis that we have achieved thus far, we know that of all the things we know to exist; there are those things that exist with form, and those things that exist without form. And we also know that both: those things with form and those things without form, also have properties that generate any number of consequences and results (which again can exist with form and without) that can also either contribute to motion or in some instances actually create motion. But on the issue of which definitively came first, motion or matter, the best that logic, knowledge and reason can tell us is: that the two are symbiotic that the one cannot exist without the other. What logic, knowledge and reason can tell us so far, is that the symbiotic existence of matter and motion transcends down to (at least) the atomic level (and it may go beyond) but that at the atomic level motion exists as part of a complete whole, a stable atom. A form of existence where particles (of some nature), co-exist in equilibrium (and harmony) with properties and powers of motion itself. And as we transcend into the subatomic it may be that the perceived distinguishing characteristics between particles of any kind, and motion itself; may become conceptually indistinguishable as existence beyond quarks may demonstrate that where the bonding of matter and motion occurs at the "strong force" level; that, at that level, new and undiscovered realities may actually take place. Realities where not only does everything aspire still further towards compatibility and harmony; but that within a "strong force" continuum, everything (still well beyond our current level of comprehension) becomes timeless, or: "all without time."

Now at this level of furthest extrapolation, we have now once again surpassed what we know; and have entered the world of potential but uncertain logical hypothesis. Why uncertain, because with each extrapolation the certainty of the assumptions being relied upon become more theoretical than proven; and therefore less reliable as knowledge. So let's go back to what we have established as proven knowledge. And what knowledge we do know, is that evolution has taken place in any number of ways throughout all forms of life, and that it is ongoing; occurring currently right now in front of us, and that evolution is observable for all to see, for anyone (or anything) that wishes to pay attention.

Now this is an important fact, why because we have already surmised that species by species, the individual evolution of each species has been on such a visible display that we cannot rule out when individual species are (or become) aware that they are in fact being observed. Now clearly we do know that most species of prey are astutely aware that they are under constant observation from a vast array of predators which lie in wait; but such an instinctive/ cerebral consciousness for survival does not necessarily equate to comprehending when they are being observed only for intellectual purposes; however we do know that many species have the acuity to understand when they are being observed either by non threatening (other) species (such as zebras and hippo's); or when (dealing with predator species) whereby the prey species will take the time (a split second or two) to make themselves sufficiently aware of their surroundings so that they can (based on proximity, distance and speed) engage in an instinctive appreciation/ calculation, for the immediacy of any potential threat. Such cognitive appraisals therefore should give us cause not to underestimate the degree of intellect or cognition that might be at play (for any given species at any given time), but even if such acuity is no more than just leaned instinctive behaviour, the probability is that such awareness (demonstrated or not) of being under observation in their natural habitat, exists as an ingrained evolved cerebral attribute. An evolved attribute of consciousness, where the precise intellectual appreciation of being observed is itself not of immediate concern; that is, until a perceived threat is recognized, and then (at a minimum) the transition (from at least the subconscious) to the conscious becomes instantaneous.

But that having been said, such should not suggest that given the open display of any given species evolutionary development that it should be presumed that a far more sophisticated interconnectivity of the individual species (instinctive/ cerebral) advancements are incapable of having an indiscernible connection in the aggregate. In fact we know that the evolutionary developments of many species impact the evolutionary results of other species when those species are in direct contact with one another. And this impact can be readily seen even when the influence takes place between two completely divergent life forms such as flora and fauna through the chameleon's ability to mimic color; or between humans and birds as parrots famously mimic our ability to speak. So to presume that each species have evolved in the way that they have, independent of the evolutionary development of the other species, disassociated from each other, is to assume a lack of a sense for synchronicity despite all the given harmonious symmetry that other forces have demonstrated through their own respective evolutionary achievements. Clearly, at the very least, the animals are watching; the question is how deep does their cerebral understanding go? But maybe far more important than each species cerebral observations might be; is the question of whether other earthly species have empyreal connections that allow their abilities to sense and "harmonize" themselves with the other invisible forces of the earth. For instance we know that both the sea tortoise and the monarch butterfly (to name just two species) use the earth's magnetic fields to navigate transoceanic and transcontinental migration patterns respectively. And though one is by sea and the other is by flight, neither ability is cerebral in the form of intellectual memory, yet such an ability means that there exists other forms of communication between our earth and (maybe untold) species which exists that transcends the existence of any given consciousness. And that raises the possibility that the forces of the earth are connecting with various species of life in ways that support the view that life here on earth, either by land, air, or sea; might be using its various different properties to assist in the struggle of evolution towards maybe an even greater level of interconnected cognition. And yes (it's now been finally said): we cannot rule out that, the earth might (not only) be trying to communicate with us; but that it has already successfully done so (communicated with some of us, by species) for thousands of years.

And the logic for this possibility is that apart from the examples demonstrated above is the reasoning that various species either: are actually aware of such capabilities through magnetoreception, and that it is therefore something to be evolved into, as in the cases of the tortoise and butterfly; or they have otherwise cognitive abilities that transcend to their very being, which requires no understanding, (a reception if you will) that requires no more understanding than a human needs an understanding to receive the warmth and nourishment (vitamin "D" for instance) from the sun. But the point that needs to be understood here is that the existence of any (and all) cognitive abilities are only being attributed to the species in each case; but what if connectivity is not itself one sided but rather exists in some form of symbiotic connection between that which enjoys cognition (with some semblance of consciousness) and that which holds transformative properties and those properties that emanate from our earth itself. Could it be prematurely presumptuous for us to assume that the existence of cognition (between species and the more sedentary) must be one sided. What if, at the very least, cognition can extend to both? Or more immediately, what if the cognition of any given species can be harmonized with the existence of the (transformative) forces in which that species comes into contact, and thereby allowing the two (on some empyreal level) to meld? What if a species with an active and receptive cognition were to make direct contact to that which is sedentary, but also transformative; could there be a resultant bond? And what if that which has cerebral cognition (even consciousness), and that which has transformative cognitive properties, could the two be attracted to one another, and if connected could there be some sort of bond? And could the physical contact of one cognition, when making contact with the cognitive properties of another; could such contact create the potential for a base form (of some sort) of cognitive continuum. On this the oceanic tortoise and the transcontinental monarch butterfly are directly on point.

Also on this, the natural equilibrium, that the earth has already achieved for itself, through the development of its own natural state, is itself consistent with such a balance of power, that the earth has already achieved for itself; thus far. Moreover, what if the equilibrium that (that we know of) has in fact already been pursued (in forms of matter), from the subatomic to the existence of our most fragile of ecosystems; are not such things the formulaic precursor to an ongoing evolutionary process towards forming a more perfect harmony of all things? Would not the furtherance of an evolutionary cognitive continuum between cognition (of any sort/ but especially the conscious), and the seemingly sedentary, but transformative; would not such a further evolutionary stabilization (in pursuit of an even greater harmony), for all things, not be: … the next logical step? And collectively, has not the arrival of human consciousness (because we have now evolved to the point where we have the capacity through deductive reasoning combined with the aided attribute of compassion) to intellectually care beyond ourselves; been exactly that next perfect solution in the creation of achieving a developing consciousness. An evolutionary solution, a consciousness that is capable of not only continuously evolving, but a consciousness that can intellectually choose to assist in its own development. Is it not completely logical that the very consciousness of mankind with all of its latest developments should be exactly (at least one version) of what would be needed as the precursor to a consciousness that is trying to manifest itself, beyond the aggregate? Moreover, must science be perceived to take the narrow view that our "middle" arrival (with our developed ability of consciousness/ in conjunction with our other attributes) has been just something we alone manufactured for ourselves, and that there is no aspiring collective force moving everything symbiotically towards an even greater equilibrium for everything. Or to put it another way, does it make sense that such solo activity by each species is to be seen as just that; or is it more rational to understand that there is in fact a more overall synchronized collectivity at play?

Light

So without a doubt; what we do know is that life is watching itself; and that life is observing the evolution of itself, when the opportunity to make those observations presents itself. How (and why) do we know this, because this is exactly what we are doing, right now. This very discussion, this analysis into the purpose of life, is our earth looking upon itself with the consciousness that it currently has. And yes there are other species who are developing other methods of observation and communications from different environments (such as through the sonar activities of the dolphins, or through the genetic seasonal migration and spawning activities of any number of varied species); but clearly the most predominant method of most species to monitor life, "observing" itself; is predominantly conducted through the wavelength transmissions of light and the reception (and the ongoing perfection) of optical sight.

So naturally, we now know (almost instinctively) that light and sight go hand in hand. But it is important sometimes to remind ourselves that not only, did naturally one exist before the other; but that of greater importance, is the reminder that, the one did indeed actually create the other. Light bathed the earth; from the earth's, very creation. And organic sight is a direct response to that reality. And although the light for the earth predominantly comes from the solar transmissions from our sun, if there were however no sun, the earth itself would not be completely without light because the earth would still be in receipt of the light of all the other stars within our galaxy (and beyond). And therefore light is a universal constant. So what is light? Well, light is a thing with form, it is the visible wavelength of a spectrum of electromagnetic radiation that can be seen with the naked eye. The spectrum itself however, also has wavelengths that go far beyond that which is visible and includes wave lengths that include *Gamma* rays and X rays, and long distant radio waves. Again, of the wavelengths that are visible that we perceive as light, are those wavelengths that give us our color spectrum and that which lies within; and this is what we regard as "white light;" and white light is transparent daylight. It is the light that appears white, to the retina; and as a wavelength, its properties grant for us the power of optical neurological imagery recreation. But also, it cerebrally gives us the power of "observation." Consequently, light and sight, through the empowerment of observation; can best be regarded as the most advanced form of cerebral evolution as it exists as a neurological form of communication.

So since we know that sight evolved out of its exposure to light; and we now know that light through its evolutionary creation of vision created the cerebral power of observation. And from there, with observation (and in conjunction with other forms of sensory communication) came the powers of deduction, reasoning and conceptualization. Thus a rapid process of cerebral advancement became exponential as greater observation lead to greater deductive reasoning, and further cerebral reasoning; then gave us the understanding of: cause and effect. And this in turn further enhanced our confidence to conceptualize that which couldn't be immediately observed, where it can now be maintained that the human ability to think and perceive of things in the abstract; is indeed now the most distinguishing feature between us and the other species of this earth. So much so, that it is has now surpassed in significance our other evolutionary developments of the anatomical adaptation of the opposable thumb, the elongation of our thumb, and our ultimate ability to adopt, adapt and manipulate tools.

So light and sight is a form of communication, but as the two words independently suggest, there are two things at play with this form of communication; one is the existence of light with all that it includes and can convey, and; second is the ability (and awareness) to receive that which light has to offer and present, in ways that allow us to understand, what we actually see.

And these two things of course must be considered separately, why: because, as demonstrated earlier with our parable of the "fallen tree," the existence and the consequences of one, does not necessarily make an impression (full, partial or otherwise) on us. Put simply, light may be conveying properties that we have yet to detect or understand. And this means that: the full extent of what light actually has to offer, may not (or has not) been received or understood by us; because we were/are incapable of receiving or comprehending what it is that light actually has to offer. Light's full properties and capabilities may well be, still unknown to us despite the fact those properties are actually here, there and everywhere; and always have been.

So with this realization, we go back to what we do know; and what we do know is that from a strictly visible perspective, the vast majority of species when they turn to the light of the sun, they are aware (in one form or another) of the light that it bestows. We also know that of those species who use the light of the sun for their own purposes, such species are also acutely aware that: that same light also equally exposes them to whomever (or whatever) might be watching them. And again, although we do not know how far such a cerebral appreciation goes; and it may be nothing more than an instinctive survival awareness (and of course such an awareness would necessarily vary from species to species) but from this awareness we know that light does transfer (even at this rudimentary level) visual information that allows even the most modest of cognitions to assess the level of threat within their surroundings. And as straight forward and as simplistic as this may appear, this awareness is important to recognize that this visual communication is an actual form of cerebral communication that is taking place. A vast communication of what the transmitted light conveys in contrast to what information our neurological brain can receive, decipher and process. For instance, we should be careful not to trivialize visible information just because daylight accurately communicates, what later here on earth can be further enhanced and verified beyond sight. For instance, when a member of a species herd, such as a baby zebra, witnesses one of its own being attacked by a crocodile while attending a watering hole, the subsequent loss of one of their own to the carnage that follows such violence and loss, allows for a tangible measurement of what has actually just happened. Similarly, physical contact through touch allows within a species for a concept of physical measurement and assessment, through such things as size, weight, density and movement; which (once touched and/or tasted) can contribute to the verification of the significance of what was originally understood only through the single dimension of simply being observed.

But when something can "only" (through distance) ever be observed, (when the reality of subsequent verification through direct contact is to be deemed impossible) the amount of the data to then be collected, retrieved and conveyed through light and sight, can trigger an ever increasing cascade of understanding that itself can create an evolution of further understanding because the intellect of the conscious observing recipient when left with no other means of assessment (through the evolution of consciousness) becomes hardwired to expand upon such neurological visual knowledge. Or to put it more directly, when we have no other means of assessing something exccpt through observation, our intellect tells us to look more closely. Because the more we watch, the more we learn, and then, the more we are apt to see. So consequently, the cerebral intelligence of the recipient becomes important because as it achieves higher levels of sophistication, the data it receives can then allow the recipient observer to calculate other such things as mass, velocity and trajectory; all based on the knowledge of the observer's ability to process this (visual) information. It is the heightened deductive reasoning based on observation that then becomes evolutionary, it is the deductive (scientific) analysis behind such mythical characters as Sherlock Holmes whose 19th, century fictional aptitude could now only be pondered to have become astronomical, if such a character should have been presented with the benefits of 21st century collaborations with NASA, and the advanced technologies of the Hubble and Webb telescopes. So light is itself, not only a form of communication, but is best understood as an expanding form of communication. And this possibility suggests that "light" itself may be akin to the universality of other such infinite things (as symmetry) which (light as a physical thing) may connect all things on an intellectual basis which creates a language where the recognition of facts, characteristics and properties, can contribute to the intellectual deductive evolution of man, almost as if it were a form of cerebral dialogue.

And so the next question that must follow is what does this possibly reveal about intellectual capabilities and "intellectual realities" such as "logic" and "curiosity?" Could "logic" therefore be revealed and proven to be an eternal thing without form, that serves as a step ladder or a stimulus for the advancement of thought, which then serves as a catalyst for instigating further cerebral evolution? And similarly, could this also mean that something as primal as "curiosity" should not therefore be under estimated, as it may actually be a more significant thing, as a potential cerebral physiological response (much like hunger) to stimulate still further intellectual activity.

So, just as a zebra may assess the danger of a distant predator and measure any imminent danger based on a rough assessment of speed and distance to maintain its survival; so too can the color of a celestial body reveal to an astrophysicist whether a particular star is getting closer or not. And again, when given a sufficient level of sophistication for a range of data to be considered, analyzed and understood, the fact that there might be an infinite amount of telescopic data that might tell us ever more about any celestial body (through a highly advanced understanding of mathematics); such an infinite revelation of a distant celestial body's unique particulars, might not only be considered (when coupled at an extremely high level of sophistication for symmetry); a base form of communication unto itself; but even more importantly it might be an actual stimulus, or an actual "invitation" for a recipient observer, to take the next step (for itself) in cerebral evolution. Therefore, much like a mother on her knees would coax a child at arm's length to take its next independent step; light is undoubtedly a proponent for evolutionary cerebral advancement whether it comes from other celestial bodies as an inducement, or from across a savannah watering hole as a precautionary warning. Such becomes the power of light.

And so why then do we flatter ourselves when we discover things here on earth when such things are presented to us by the activities of other species, or moreover are paraded before us through the light of the night's sky. In both instances we commend ourselves because we choose to compliment ourselves because of the deductive aptitude that we have accomplished when time and space may have made such findings apparent, millennia ago; and we as intellectual recipients, have just now awakened to it. Mankind has been conscious for thousands of years, so can we claim that in coming to these discoveries of late, can it really be said that we have done so, ... by doing our best? We cannot assume that time for us is infinite. Before us, there were at least five life extinction events that we know of. So again; can we really claim that for thousands of years we have aspired to do our part in order to evolve?

What we have to recognize is that: "discovery" and "revelation" are the same thing, the only difference is which side of the electromagnetic epiphany, is the realization coming from? And since matter and motion might best be understood to be a continuum, where it appears that there is always either motion or "strong forces" within the existence of any given particular particle; than the origin and the evolution of life may well rest with the existence of these instigating (and driving) properties; regardless of whether we choose to recognize such forces as "properties /or a thing that has any sort of "cognition." The question about the origins of life therefore, may be dependent upon the very old (subjective) question of: when does physics actually become chemistry? Or if we are willing to delve deeper into the subatomic (as postulated above): and ask the question when/or do: the forces of property and motion, become (by any definition) potentially cognitive? And although, it makes complete sense to argue that the origins of life, seems to all revolve around the existence of an electrical charge, and this seems to be true; does not an electrical charge itself come from motion, or does motion create an electrical charge,? So we come back to "the beginning;" where we are driven (by evolution) to ask, which came first: "motion" or an "electrical charge." Or are they both, so close to being simultaneous that they are just indistinguishable?

CHAPTER FOUR:
THE PURPOSE OF HUMANITY

So what we have established thus far in our analysis to determine the purpose of life is that in posing this question it became apparent that we didn't have a working understanding of what life has been, and is. It was understood that there were also component parts to the question itself as it was obvious that "life," here on earth; is comprised of more than just humans as a species, and that we humans (as we are the analysts in this enquiry), would have to decide how we were going to conduct our enquiry since we too were part of the equation that was to be observed. So before such an analysis could commence, it became apparent that a recognition would have to take place where the scope or definition of life would itself have to be sufficiently expansive so that any purpose that was to be surmised, would be potentially inclusive enough so as to be relatively accurate and not prematurely flawed through a presumptive act of omission. And to create this most potentially inclusive definition, it became logically incumbent upon this enquiry, to then try to grasp the potential origins of life so as to be as inclusive as possible for all "life" that may have subsequently evolved or transpired.

So: as for our foundational definition of "life, the working presumption that appears to be the most inclusive is that: it must include all things that exist with form or without form; that may exist organically, inorganically, impregnably or as partial code; and that it may possess the properties of either internal atomic movement or strong force properties. And if from this: such a definition may appear to be too inclusive so as to leave very little out; well that's just the point. What is to be understood from such a definition is that it is a working definition that is infinitely transcendent, it is a working definition that allows us to delve ever deeper into the subatomic while simultaneously looking out to the cosmos for all the information it has to offer. It is a definition that expects the unexpected, and it doesn't presume that there is a known starting point. On the contrary, this definition dispenses with the distinction between physics and chemistry, in fact the opposite proposition is proposed here namely: that the origins of life traverse those very distinctions between the inorganic and the organic realities. Furthermore, by disregarding the distinction between physics and chemistry, this definition adopts that science anticipate that by maintaining that life exist in a vessel of physical form organic or inorganic, that such a physical requirement is flawed because it precludes the possibility of life (or its properties) existing as an invisible collective. This most expansive definition anticipates that there may be the existence of life that exists (and is best understood) in the aggregate, existing more on properties without form, than existing as life dependent upon physical matter. In short, the search for "life" as a concept might better be served by science by allowing for such collective realities as those which exist in the same realities of such things as murmurations, where a flock of starlings can simultaneously change flight direction in unison; thus creating the a visual equivalency of a symphony in the sky, because what we do know from the formations that this symphony presents, is that it is meant to be seen (maybe even, for us); while equally creating at the same moment (but only just for an instant/ a millisecond, for other observers), a precise replica of a monstrously large bird of prey, as a defensive mechanism.

So although it makes sense that we look for the purpose of life by looking at the earliest visible origins of life, such a preoccupation may be limited/and limiting, because it assumes that physical matter gave birth to life; but since it is these very same things (that we regard) that make "life" special; such things as those without form, such as consciousness, intellect, and awareness that may be connected to other such eternal things as: logic, symmetry, equilibrium and precision. Because what if we are to realize down the road that such empyreal things as symmetry, light and logic are themselves the precursors for the evolution of consciousness; then maybe we would be better off to start our search with those achievements in mind; as they may hold the very essence of what life has actually always been all about; since the very beginning.

And it is here where the issue of "cognition" once again reappears as a constant when the word itself is understood best not as a form of consciousness, but rather as a foundational precursor upon which a developing form of consciousness may follow. "Cognition" in this sense, broadened to include "cognitive properties," can and should be used as an active form of terminology, an invitation to recognize that which has "driving properties" of either "bonding" or "movement;" which may signify a force/ or forces which may hold the powers for a "thing" to possess the power to provide for its own "continuation;" or, more assertively hold a potential power for adaptation. And it need not mean more, but it might. And that is the point. With an expanded definition, an understanding of a base form of "cognition" can encapsulate unknown driving forces of life, that at any given time, may not meet the standards (if: in this context there are any) of consciousness; but, as words ("cognition" or "cognitive properties") may represent the existence of initial "sustaining" or "driving" forces for the later evolution of life, nonetheless. Cognition, therefore, as a word; can represent the concept that the origins of life should not be constrained by the conceptual bifurcation between organic and sedentary/inorganic matter. Cognition in this sense can therefore theoretically traverse from one to the other and therefore as a concept, the word can present itself; as a constant. Such should be both the gymnastic flexibility and the power, within the: "word".

But having said all that: what really matters is the original proposition that our search for the meaning of life should simultaneously be pursued by looking transcendently both inwardly to the subatomic, while conversely looking in the opposite direction towards our outer most cosmos. Moreover, as we stand here in the middle of time, space and evolution from a multidimensional perspective we should realize that the achievements of things without form such as consciousness, awareness, intellect, conscience, and the powers of conceptualization and creativity, …to name just a few; that these achievements are not without their connection to the purpose of life. We must remind ourselves that all these things, and more; are real. Because: we are real. And we are the most advanced part of the earth's evolutionary process, thus far. And, just because these things exist without form, or are more commonly referred to as a person's thought processes; such characterizations should not, through their daily rudimentary status, diminish the significant evolutionary adaptation of cerebral advancement that they are. Our current powers of intellect are just that, forces of immeasurable potential that heretofore, have never (both in number/by global population; or in depth/by technology) been reached before; and which in substance and content are now surpassing any of our previous heights of accomplishment exponentially; with each passing unit of time. On this, our exponential increase in capability through the development of AI would be a case on point, as we have now moved from developing and using tools for our adaptation; to being on the cusp of having our tools being able to develop tools for themselves. And our current position of enlightenment matters not for when the process of evolving cognition to consciousness began, only the reality of how, why and where; this transformative cognitive evolution came to be. But the reason, the how, the why and the where; are so significant to the meaning of life is because the more connected our life is to our planet's origins and our cosmos, the more our individual lives are best understood as an actual extension of both.

And herein lies the rub, because the more we find our existence to be an extension of our own earth and galaxy, the more our consciousness is the developed consciousness of our galaxy. Therefore, in reason and logic it can be said that the purpose of humanity is to give life and consciousness to our galaxy, the Milky Way. And why do we know this? Because, we exist. Because the physical matter that makes up our galaxy has evolved into life, and again we know this (without a doubt), because we are here, and here we are. So from, the first celestial particles (which in its unfathomable mass), formed our galaxy, to the creation of our solar system, to the evolution of the consciousness within all forms of life here on this planet; mankind and mankind's consciousness, is itself just the furthest creation of this continuum. In short, our consciousness is the consciousness of our galaxy; and it matters not whether we are one of one, or we are one of many within the galaxy; what matters is that either way, we understand through the existence of celestial consciousness (that we now know does exist, because again, here we are), is that our galaxy's consciousness: is who we are?

Celestial Consciousness

Now the reason that it is appropriate and accurate to call our consciousness; a celestial consciousness, is because it is irrefutable. Why, again because we know as a matter of fact that humanity is a reality, we exist both individually, as persons; and as a species, homo sapiens. And that as a distinct species we are however, just one of a vast array of other species, that are also of this earth So as mentioned earlier; as we engage in the enquiry of what is the meaning or purpose of life, the question remains are we entitled (based on the strength of our intellect) to restrict our enquiry on the subject matter to answers that primarily relate to the more exclusive issues of human concern? Or, as the researchers of our own study, are we somehow obligated to be inclusive of all other species, simply because through their evolutionary development, they are unable to engage in a cerebral dialogue as to whether or not they have a purpose. But maybe more importantly, is the question that lies underneath, namely that: if the many species of the earth, and in fact the earth itself; are to be included in this discussion, is their inclusion to be out of charity and benevolence, or is such an inclusion to be pursued as a necessary and absolute imperative. For, the answer to this question; will inevitably decide the other.

So for the moment, let us presume that an enquiry into the meaning and purpose of life ought to be all inclusive; then the question becomes do we make the decision to have all forms of life included in the analysis; do we make this decision out of benevolence and compassion, or is this question itself, a logical and physical imperative? And on this, the factual case for the imperative, becomes clear: and that is that the inclusion of all life (in search for the meaning and purpose of life) is indeed an imperative because any exclusion of all the possible options and information attached to those other species would be to distort what life itself has already created. Science tells us that evolution: past, present and future requires that if evolution itself is to survive, as it has done in the past and up to now; then it cannot thrive if it is to be denied (through a distorted perception of purpose/ by its most cerebral participant) the chance to offer other logical alternative explanations for how things are. We must remember that we too are the voice of logic and reason and if we don't give a voice and vocabulary to all logical possibilities we effectively stifle and limit by shear definition alone that which might be actually possible. Our own evolution of DNA itself is a case on point, the double helix of our DNA genome evolved beyond the limitations of the single helix RNA genome, because to create the potential for humanity (its survival and advancement), with our DNA's current drive for life, to do this, it simply had to do …what it did . Why, because that's what it did. The fact that it has happened, as a matter of creation (distinguishing those things created / as opposed to those things which through thoughtlessness can be destroyed), means that we must leave open the possibility that the driving force behind DNA actually "aspired" to evolve into the double helix that it has become; or, that it simply had to become, what it actually became. Clearly, by the same logic, it may not have been by design or out of necessity; but it's entirely possible also that the double helix simply came into existence by chance; but to rule out any of these three possibilities is to be prejudicial and illogical. (And, yes; we know how that sounds.) But the reality is that: humanity; exists as it does today; by the benefit of the double helix. And it would be highly illogical (and there's that word again) to rule out any of these probabilities without knowing more.

So where does that leave us? Well currently, we regard humanity as the latest and most sophisticated evolutionary adaptation that our earth has accomplished thus far. But the assumption that our single species intelligence is the most advanced form of intelligence possible, is itself a rejection of the possibility that there may be an even larger collective intelligence to be discovered here on earth, developing in the aggregate. It cannot be ruled out that the powers that currently drive life, all life; have influenced (if not actually shaped) the resultant evolutionary traits where the whole becomes greater than the" sum of its parts." We don't know why this happens, but from this very fact alone, two likely possibilities become apparent. First, there are actual other outside forces and powers at play (such as the true impact of such things as symmetry and light) that we don't as yet understand; or second there is within the atomic/cellular structure properties, that allow/or anticipates the next available (or dare it be said: logical) step. And whether the influences of outside forces stand as evidence that there is in fact a greater "collectivity;" or whether internal (possibly atomic properties) are at play; creating a particular (possibly cognitive) outcome; either way, earthly consciousness has itself become a fact; and human consciousness (appears right now) to be the most advanced singular form of consciousness that has evolved thus far. So yes, evolution has the power to make something greater than the sum of its parts, and so for us in our search for the meaning of life; for us to thus look at life predominantly, or solely, from a human perspective or agenda; is to deny the existence of powers and forces that we can only presently speculate upon (if we can even imagine them, at all). So the only logical conclusion that can be reached is that in attempting to analyze the purpose of life; not only are all other species to be included in this analysis, but also all other such things that exist without form (such as logic, symmetry, etc.) should be worthy of inclusion; because, in the alternative, without such an approach the achievement of a more perfect "truth" (through knowledge) can be prematurely (through deliberate omission) effectively denied.

So the conclusion that life must be all inclusive, to discover the most comprehensive meaning of life: creates the analytical advantage of approaching the task from a multitude of perspectives aided by the cumulative data that comes from our best research of the achievements of each species (as we have come to know them). So the regenerative capabilities of lobsters to regenerate amputated appendages such as complete claws, antennae and walking legs; or the peregrine falcon that can dive bomb and can control its descent at over 240 miles per hour; or the vast array of snakes and reptiles that possess the abilities of infrared sight; or (much to our domestic distain) the fact that urban pigeons can perceive shades of colour vastly more discernible than our own (fashion divas), including ultraviolet light; all of these abilities (and a multitude more) are the creation of life's many specie (d) powers of evolution. And yet despite such diversity, and the particularization of specific skill sets within, because of mankind's particular development of cerebral abilities, mankind through our own preoccupation with our own affairs, routinely demonstrates a willingness and tendency to separate ourselves from the rest of our earthly residents, especially when a perceived human self interest or priority, presents itself.

Mutations Are Real

Now to our credit, what we should be observing and searching for "in everything and everywhere," and frequently fail to do so in too many other evolutionary instances; is that when it comes to observing and tracking the evolutionary traits within humans, our success rate is somewhat better. But this is no doubt in large part attributable to our ability to complain; and because it is in our own general best self-interest to recognize in others, that which might eventually effect all of us, both individually; and as a collective species. … And of course then there is always that other persistent issue which is both an accelerant and an inhibitor, namely: the driving force …behind/and about "money."

But; that said: everyday science and medicine are barraged however with newly discovered cases of genetic aberrations or mutations that lead to devastating and debilitating illnesses and diseases that are often chronic and frequently terminal. So common are the frequency of these genetic breakdowns that through mass media, the frequency of these genetic breakdowns have now made many diseases household names, even if the genetic cause may not be so commonly known. Conditions or diseases such as: sickle-cell anemia, hemophilia, cardiac heart disease; and the more common treatments in cancerous melanoma, retina repair, etc.; all are conditions and treatments that are connected to genetic mutations that through one form of gene therapy or another, including genetic replacement and designed genetic (reparation) mutations; have all benefitted from genetic treatments and procedures with varying (to promising) clinical and practical results. Such are the advances made in medical science through genetic decoding coupled with the research gains made in the study and application of stem cell research. And with all these medical therapies at play, what mankind has demonstrated is that we can contribute to the genetics of our species as we repair or minimize the genetic impact that lead to an onset of illness or disease that follows any particularly (flawed) genetic trait. And although such genetic techniques and therapies may only ease the symptoms, complications or positively provide some relief; in many cases, our intervention still leaves us regrettably, without many desirable cures.

But in the search for cures, and especially genetic cures, when our (i) search for life and (ii) our search for the purpose of life, is then assisted by (iii) the evolution of life itself; these three can reveal (from time to time) a convergence of evolution, consciousness and purpose, where a harmonious collaboration can take place (for all to see), demonstrating how miraculous the possibilities of life could be; especially when all three "influencers" converge. And in the field of genetics, we have the exceptional discovery of Jo Cameron, a Scottish national, who in 2019 was discovered to have an extremely rare gene mutation that impedes her ability to feel pain, in fact she has almost felt no pain for her entire life. Her mutation in question is specifically with the FAAH gene which was originally ascribed to an area of human genetics which was regarded as "junk" genes within our DNA code, simply because this area of genetic mapping was initially thought to be rather redundant. But now however, due in large measure to this discovery, this particular region of DNA mapping is now to be viewed as holding much of the information that actually regulates the gene coding function itself. Such a discovery as this, is what happens when, consciousness, purpose and evolution itself, converge as a collaborative process working in unison. Put simply, when consciousness, a sense of purpose (through clinical research), and evolution itself converge; more rapid progress is made. And this is precisely the type of progress that the earth has made through its own evolutionary achievements through the accomplishment of consciousness, through its many varied species that resulted out of its creation of organic life. And now, through the power of consciousness; the evolution of what that consciousness is used for (ergo: purpose), can now make the rate of that change (consciously) exponential. So yes: we have now come to the conclusion that our galaxy itself has its own purpose, and that purpose is to realize its own consciousness through the creation of us, and yes: that means all of us, the earth itself, we the human race, and all species alike. And how do we know that the galaxy has now achieved this consciousness, we know this because this is precisely what it has done; again, the galaxy has created the earth and consciousness for itself, through us. We are the consciousness of the galaxy; and the most immediate way to fail that consciousness, is simply to fail to realize this irrefutable fact.

Now of course the therapeutic potential of the discovery of the FAAH- OUT mutation has been/ and will be immeasurable, but as important as those results will be, the fact that evolution through consciousness has shown us that life without pain is actually possible, our consciousness can now search for other beneficial mutations that will not only similarly benefit humanity; but will allow humanity to possibly assist in the shaping of other genetic evolutions, in ourselves and all other species. (And yes; that is a little bit scary.) But scary or not, humanity; with consciousness and knowledge, can now play a role through "gene therapy and repair," that can aid evolution in its strides forward. And clearly, the control of chronic pain (if it can be achieved) has certainly got to be part of any conscious choice for a continued pursuit for the purpose of life.

Consciousness: Part I / Humility

But, why? Why does gaining genetic control over our pain receptors have to be part of our overall pursuit for the purpose of life? What is the driving force that makes this such an evolutionary thing to do? Well clearly the achievement of consciousness is so much more than just the acquisition of just another physical ability, such as the development of the opposable thumb. In fact it is a question of some significance of which came first, did the appearance of consciousness give birth to other abilities such as the versatile thumb, or did the abilities of other such anatomical advancements such as the thumb, awaken and broaden humanity's consciousness? It may be that the two (and others) were just symbiotic. And what happened was that: whichever came first, consciousness or anatomical abilities, the first beginnings of either one, made the earliest advancements of the others, more promising. And it is this interconnectivity of each evolutionary development, that with each tiny step forward (the more atomic or microscopic the transformation) of each development, the closer in time do the developments within consciousness and the actual anatomical developments seem to become, in association with one another; to the point where the developments in consciousness and (microscopic) anatomy, begin to appear as if they are instantaneous in development of one to the other. And this tight proximity in time, substantiates the proposition of a possible innate sense of cellular cognition as the physiological cell structure attempts to accomplish ultimately what the cerebral intellect is trying to do. Or conversely, an increase in anatomical reach and dexterity stimulates a conscious realization that something new is possible. So in either case, (in its most rudimentary state), this just means that where a primate (such as a capuchin monkey) either experiences an increase in precision dexterity and realizes that it can now do more (hence, anatomy leading consciousness); or it successfully conceptualizes that it needs to reach further to reach (let us say) water through the assistance of an absorption stick, (so that a stick can act as a type of elongated sponge) to capture water that is out of reach (thereby having consciousness take the lead in anatomical dexterity). And the point here that needs to be made: is that the closer in time that this symbiotic relationship takes place between consciousness and microscopic cellular adaptations, the more credit the one function/or adaptation has to give to the other for being the actual "brains" for the adaptation itself.

In fact it all just does seem to fit so remarkably well; together, for as in the case of the capuchin primate, the originating hunger and thirst of the primate, the appeal of available but inaccessible water, the water's preserved capabilities of an encasing receptacle (such as a tree). Then the recognition for the need of an extension device, a stick; and with the stick, the actual realization that the water can thereby be retrieved. Furthermore, the realization that the volume of water held by the stick can be increased by chewing on the dipping end to make the end function more as a sponge. And the symmetrical equation of removing one organic supportive source from its protective outer layers; all of this presents a harmony of physical, conscious and empyreal activity that appears to be consistent with a broader assessment of a collective whole, symbiotically working together to provide for the very existence and the survival of all.

And why is this all relevant, it matters because the driving force in either the case of mutated pain receptors for homo sapiens, and/ or the primate that creates an absorption stick, demonstrates that evolution can come from either: within a DNA instigated mutation (by design or by accident) where (through advanced consciousness) Ms. Camden was able to report on her adaptation, which turns out to beneficial for all of humanity; or it can come from the most unassuming consciousness of a primate that attempts to achieve a little bit more (water), with the simple use of an absorption stick. And why does this matter; because both (the DNA mutation obstructing pain, and the tool manipulation of a stick) involve an advanced form of pre-existing cognitive/property (whether it be consciousness or cognition) whereby each species is doing its best to evolve as far as its current state of evolution will allow. DNA itself does not have consciousness, but it is capable of both transforming (and is transformative); so the question of whether its developments are by design or by accident isn't (immediately) important. What is important, is that there is sufficient "force" activity within, or circulating about, to trigger these transformations and, whether by accident or design, hopefully the beneficial transformations will survive and continue, either by being handed down genetically; or, by being discovered, as Ms. Camden's mutation was discovered first by her; and later by, allowing us (the science community) to capture its essence and existence for further research.

And as for the adaptation of the capuchin primate (the primate with the absorption stick), the cerebral adaptation for simplistic problem solving should not be underestimated, because on a more sophisticated multidimensional level, this simple development has a multitude of ramifications because first it may be a first adaptation for that species, or it stands as an extension of a previous use of tool manipulation that was mastered earlier, either way it is (within that species) a cerebral step forward. But secondly, on a far more complex basis, with us observing the capuchin activities, such observations create a paradigm where a connection is made between humanity and other similarly observed animals and primates. A connection and paradigm, which (through sight and observation) creates within us a sense of responsibilities and a sense of awareness that allows us to conceptualize the possibility (of what is actually being discussed here) which is namely that: while creating the (less helpful) appearance of superiority (for humans); by us observing the divergence between a different primate and us; what is actually being manufactured successfully, is a greater sophistication which stimulates (within us) a heightened sense of both self imposed humility and a sense of responsibility; all at the same time. So should we choose to focus our intellect and marvel on our own superiority, we can then stifle our own evolution; or through the wise choice of humility our evolution can become potentially infinite. A humility that allows us to see how small we are when viewed from a larger global life perspective; while substantiating through that recognition, a sense of life force collectivity, for all life. So with the visualization of the divergence, the contrast itself; becomes the teacher. And with such teachings, humility and a more accurate sense of self awareness of both ourselves (as individuals and as the observing analysts) inspires (for us), oddly enough, through the powers of evolution; a self realization of where humanity might actually stand within our galaxy. All this and more, because at the behest of a primate searching for water, our minds become expanded because of our observation of a (colloquial) monkey: "waving its magical wand."

This is the wonder of zoology and botany, both these and other disciplines have the power to stimulate our conscious intellect to heights of sensitivity and reflection that can only be maximized by stimulating the best analytical part of our cerebral, which is our intellect. So through the intellectual process for conceptualizing both contrast and commonality; but maybe more importantly, rather than viewing the earth as creating a global greenhouse for the benefit of humanity, we should see ourselves as part of the equation where through the cerebral process of comparison and contrast; our earth, by creating the ability of consciousness, has also empowered that consciousness to be stimulated for the creation of other such perfected powers and attributes such as the power of "curiosity." And if consciousness comes with its own instinctive need for knowledge, then curiosity is then the insatiable adjunct to that need, which seemingly never appears to be satisfied. So just as cellular existence requires thirst and hunger to facilitate the ultimate process of cellular replication; curiosity therefore; is similarly a driving force for the intellect, a hunger (attribute/ if you will) without end. And to this purpose, once again, the observable zoological and botanical evolutionary results of all other living things; to the conscious intellectual observer, these supposed less sophisticated and (in the case of plants) unconscious things ... (like the monkey) become the teacher. Not for humanity separately, but potentially for the self realization by humanity for the earth's own evolving drive to take its collective forces of cognition beyond its own current aggregate. So consequently, but "not so as to make too fine a point about it;" the earth (while working through its consciousness in the aggregate) could be pointing us to the revelation that with our advanced form of consciousness; we have been entrusted with being the representative consciousness (a "first amongst equals"/ if you wish) of all life, including all those species that can't speak for themselves.

Consciousness: Part II / Our Strengths and Gifts

So in the particular case where we, homo sapiens; have genetically developed to the point where we have already created the electron microscope and we have proven that we can travel into space. In the case of the capuchin primate not only have they recognized that a stick can be used as a dipping device for reaching water but they have also discovered that they can increase the absorption rate by first chewing on the dipping end to soften and broaden the amount of water it can hold. So what has to be recognized from such examples is that if we look at animal life as we know it; we can obviously recognize not only that almost all the land species that have developed their own form of consciousness (as they perform their daily routines) but they have done so by acquiring very much the same anatomical receptors for sight, sound, touch, taste, smell, etc. in order to service that consciousness, while each species has simultaneously perfected each receptor in their own way to uniquely accommodate their consciousness as it endeavors to provide for the survival of itself and its species in general.

Consequently, the striking similarities however in such anatomically created sensory receptors (eyes, ears, nose, etc.) and their ultimate service to a central cerebral nervous system, demonstrates overwhelmingly that cognition and consciousness are not only a constant recurring theme in evolutionary development but that these two achievements (at whatever varied level of development a species may be); that there lies within each, the potential (and possibly even the "aspiration"), to possibly through the harmonization of all, to become "more." That is, to possibly become more than to have acquired a particularly enhanced form of specialized sensory ability for any one form of our five typical receptors. On this the obvious canine distinction between wild wolves and domesticated dogs stands as a strong example of this genetic variance capability. Both wolves and domestic canines have particularly advanced forms of receptors for the acuity of smell and hearing, but (like humans) dogs have also evolved "consciously" to become "more."

Unlike their genetic cousins, the wolf; domesticated dogs have over the past 100,000 years not only evolved in their companionship with us, the homo sapiens; but their evolution has evolved in such a way so as to become the predominant genetic companions of humanity as well. On this point, a dog's DNA is 84% compatible with that of human

DNA. And although such a high degree of similarity means that at some point (close to 100 million years ago) man and canines shared a common ancestor, the high degree of similarity at 84% can be attributable in part to a high number of common gene sequencing events that both man and dog have been exposed to over their long relationship. So close, in fact, has been this relationship; that it has been surmised that a fully domesticated "border collie" has been speculated to have a potential auditory vocabulary of between one hundred and fifty to three hundred words; and the potential IQ equivalency of a (2 year old) human toddler.

That said, the above facts are not really the reasons that dogs have become referred to, and known to be as: "man's best friend." This particular quality and characterization speaks to the qualities within a dog that go far beyond its trainability and their overall utility. The characterization that makes dogs so compatible to human beings (besides their potential for obedience) is their domestic willingness to bond and display outwardly their perceptions and their emotions. Dogs clearly possess the emotions of fear and affection; and theories abound that dogs have (through their gene sequencing) evolved in synchronicity with us because they were intuitive enough to recognize that an affiliation with mankind could actually be a survival mechanism. So through companionship, with another species; mankind and canine, have been the successful beneficiaries, of each other.

And herein lies the turning of the coin, domesticated canines clearly demonstrate that life within a species can aspire to become more. But this too, is exactly what our galaxy has actually already accomplished (except on a significantly broader scale); as it has evolved/ or transformed from the inorganic, to the organic, to consciousness, and now to what we currently describe as cerebral intelligence; and with cerebral intelligence comes the powers of conceptualization and the power of free will; and of course with the power of free will, comes the ancillary powers of individual and societal choice. But to get to this point in evolution; all the species of this earth and humanity had to arrive and get to where we are presently, where in time, space and evolution, to do this (whether it be by design or by accident), we as a species (like so many others) had to become, "more." And as a species to make it this far, for homo sapiens conscious intellect was not to be enough; to arrive at our current abilities, the will to survive had to become more; it had to progressively evolve into a "will to safeguard" and that our "driving will" to safeguard, also had to expand into the capability to be able to "sympathize," "empathize," "provide for," "reach out" and "defend;" those individuals and species that were beyond our progeny. We as a species had to learn to feel "compassion" (in the abstract) towards others who were in obvious distress, pain or need. In short, to get where we are now, we had to develop the essence of a "conscience;" and with a conscience, the concept to love beyond one's "self" and beyond our own "progeny;"and we did this as a matter of evolution so as to become our own best defense for survival against famine, violence and war. Ergo the flipping of the coin, where the evolution of life through time, on one side of the coin; had us to evolve to provide for immediate survival of individual and species alike; whereas the other side of the evolutionary coin, the evolution of life has accomplished through the same passage of time, those intrinsic things (now within our galaxy/ because we are the consciousness of our galaxy) the very things that makes life actually worth "living for."

So life, has its purpose; and though we as humans exist with our cerebral abilities to appear to be at the apex of all evolutionary life, the fact is: that despite our intellectual prowess, we as a species are still just one of many, and it is in this sense that given the abundance of alternate life on earth, all which appear to have a consciousness (and an array of impressive physical achievements) of their own; it becomes presumptuous for us to assume that there isn't a more sophisticated essence of intelligence and consciousness already here on earth that may only exist (for now) in the aggregate. A cognition, possibly still in development; where the earth, that has created everything else; is now using its latest and most advanced form of life, the homo sapiens with evolved abilities of cerebral consciousness, and now with the additional abilities of compassion, empathy, kindness and deductive reasoning and purpose; to observe and process it all for the benefit of all; and thereby to discover and finish building the necessary "Wi-Fi" (as it were /organic or not) to bring it all together.

CHAPTER FIVE: THE BATTLE BETWEEN GOOD AND EVIL

So what can be deduced from our analysis thus far: in trying to ascertain the purpose of life, is that there can be seen a logical progression in the evolution of life, ... when the concept of evolution itself, is viewed as a collective whole. And clearly, individually, each species has had to do a great deal of development on each of their own, in isolation, as each species would create a vessel body upon which that vessel body could best adapt to its environment. And herein lies probably the best explanation as to why mankind has had such a hard time in visualizing life as a collective whole, because there are at least two phases of isolation for life that conscious life must endure as a matter of harsh reality. The first takes place with the initial imperative of creating an individual vessel, a body which can be born and that can best provide for the most immediate survival of an individual conscious life through such things as breathing, eating, drinking, sleeping, etc. And then the second, is the isolation that the collective species itself experiences, as its size and numbers must then compete with other species for the necessary space, resources and expansion that such a species (through growth) thereby creates. So consequently, a thriving species must be able to exist with either an ever expansive domain of necessary resources, or it must reach a harmonious balance (an equilibrium if you will), with its' surroundings; whereby its growth in size becomes compatible (and sustainable), within those surroundings. And of course this economic reality thereby limits the size of any species both in the individual body and in the actual number of the species itself; and once that apex is reached; and should that equilibrium of species consumption/ to environment become breached, then the species must either migrate to fulfill its needs, or it must evolve in a manner that is less nourishing than before. And should the conditions become sufficiently harsh and bare that ultimately suffocation, starvation or dehydration become manifest; then death and extinction become an imminent reality for both the individual and species alike. Such are the harsh and seemingly "cruel" ecological realities for those who experience such hardships (and for those who observe) such desperate economies of biological life.

So as analysts and observers, we can recognize that biological and/or organic life, can only exist by virtue of their physical needs, demands and imperatives being at least fulfilled to some extent. And as observers, we can concede that: with physical existence, the bodies that envelop "life's existence", are fragile and perishable and therefore, are in need of constant replenishment. Such replenishment, that if exhausted or not maintained, cause deprivation, loss and death to follow. For us as observers, when we watch life become imperiled by starvation, thirst or dehydration, it becomes our evolutionary responsibility to at least recognize that with such deprivation; that the inevitable realities of physical suffering and hardship (for life that is conscious), at this point begins.

That said however, life as it exists in all forms of animal species; such species universally do possess consciousness and (as demonstrated above) also often display a capacity for feelings, attachments and emotions, as do we humans. And with this, consciousness comes (as expressed earlier) an awareness where with the physical need, comes with the earliest manifestations of hunger, thirst, and exhaustion; and, if left unresolved, these needs escalate to the painful realities of starvation, parchment /and dehydration; and ultimately to illness, disease, organ failure and collapse.

(Now conversely, and for the benefit of contrast, it is contended that when plant life is deprived of its needs and imperatives that through such sustained deprivation such species are apt to wither and consequently die; but because plant life is deemed here to only possess cognition, but without consciousness or sensation; therefore with such deprivation of plant life's imperatives, plant life's mortality, comes without any experience of pain or suffering.)

So from the above, with respect specifically to all forms of animal life; a physical duality is created where, with the benefits of life; there is created two physical realities of adaptation in both consciousness and anatomy, as a direct result of evolution. At the apex of evolution is the accomplishment of consciousness for all animal type species, despite the fact that the level of consciousness and degree of sophistication will vary immensely from species to species. That said however, the reason that consciousness is at the apex of evolution is because regardless of any particular level of development in anatomy or consciousness; consciousness itself, allows each species to strive for more. And as expressed earlier, the evolutionary adaptations that each species will ultimately achieve, will result either out of an environmental short fall, thus creating a need for more; or, out of consciousness, where there will be a realization that there can be more. And from these two dynamics, the resultant duality of adaptations will take place both in either anatomy or in consciousness; where (as expressed earlier) in the early stages it may be difficult (even/or especially through the process of natural selection) to discern which evolutionary change came first, an anatomical adjustment or a conscious awakening? And of course, at the most early stages of any organic evolutionary change, one can expect the anatomical initial adjustments to be almost microscopic, while the conscious adaptations may similarly be indiscernible because the source for any initial revelation may come from any aspect of the species life. Conscious adaptation could come from something the species sees (or has seen), or experienced, from any of its other sensory receptors; or it could be genetically inherited; and of course depending on the species in question, we might never know if whether a species, in contributing to an evolutionary development, may have done so (at its earliest stage) as a result of (dare it be said) an "epiphany."

But regardless of whether the adaptation comes initially from anatomical evolution as a species reaches for more; or consciously, by conceptualizing and then choosing to reach for more, such an evolutionary conscious development, for mankind has now far outpaced our most accomplished anatomical advancements in the past (such as the opposable thumb) so much so that now, our consciousness, continues to expand at an almost exponential rate. And with this evolution or expansion, our adaptations and achievements have now become exponential through technologies such as AI, Quantum Computing, and our collective creation of NASA; but because what has happened, has already happened; namely that humanity exists today as it does, because we have been endowed with not only consciousness but the best parts of consciousness, which include: intellect, conceptualization, abstract thought, an appreciation of logic and deductive reasoning; but also just as important, conscience, compassion, empathy, responsibility, fairness, equity, justice, loyalty, love (and this is to name just a few).

But unfortunately, with these strengths that were born out of a harsh physical existence and the achievement of consciousness (and all that has followed); the harshness and hardships of physical existence have also come, and through the powers of a twisted consciousness, have also been born the cancerous byproducts of physical existence, the worst of which are the realities of pain and suffering that comes from that which threatens life individually and collectively. And the threats that threaten us most directly and immediately, as individuals; can come from at least three possible sources. First, from the internal breakdown of our own physical, individual existence, where from the rigors of biological and anatomical needs, our cellular structure and replication process cannot be maintained due to an immediate internal threat from such things as cellular breakdown, organ failure, illness and disease. Second, as individuals we can be immediately threatened by other individuals who succumb to their own cancerous byproducts of consciousness which twist their individual intellect to act upon, their own conscious negativities of vanity, ego, envy, cruelty, jealousy, greed, hate, and indifference, and then by targeting those feelings, typically through acts of violence, towards others; thereby threatening any other individual immediately and deliberately, at any given time. And thirdly, direct threats of imminent harm, can be placed on the individual (and any given collective) through any deliberate collective manifestation and weaponization, of such things as: poverty, displacement, famine, thirst, disease and most poignantly through acts of war.

Such are the evils of destructive forces. When; first, we stand by and allow such things as illness, disease and accidents to run their course when we have the compassion, knowledge and knowhow to stop it. Second, when we allow the cancerous byproducts of physical existence and consciousness, again through such extensions as vanity, ego, envy, cruelty, jealousy, greed, hate, apathy, want and ignorance; to rule our individual intellect. And third; when collectively, we act on our individual cancerous flaws (as named in the first two); through such collective aberrations such as dogma, tribalism or extremism, to weaponize mob mentality into a false philosophical pretense for violence and war. These are the evils that grow in the vacuum of one's consciousness when the "good will" of man (made available through consciousness) is left vacant, and this vacuum can be filled quickly with such conscious negativities, when such emotions are stimulated though the creation of anger, frustration, hate and despair, brought about by the physically anguishing sensations of deprivation, isolation, exploitation, loss or hardship.

The Evolved Attributes of the "Goodwill of Man"

So when the "strong force" evolutionary advances of consciousness, best characterized as "the goodwill of man," come into direct contact with the pending immediacy of a recognizable catastrophic loss; physical survival takes over in that moment, as the proximity of that which is in danger of being lost, challenges the depth and strength of the observing consciousness, to act on the advances for "selflessness" that have become potentially ingrained into the "very being" of that individual. This is what occurs when an individual recognizes the imminent danger of losing life, limb, or a child. The observer's reactions to an imminent threat are based on a conscious (but in this case an ingrained subconscious) correlation between the pending physical loss that is best described as instinctive; in contrast to the degree of strength within the consciousness for "selflessness," to combat intellectually the immediate fear of personal physical harm to one's self when faced with the need to help others. So where a mother will run into a fire to save her child, such an extension of selflessness is quite frequently instinctive, because of the child's proximity to the mother's self (and although such a selfless act is easily understood / it is still an extension of selflessness nonetheless). But, in those instances when another person may well do the same, but as a stranger, such similar actions may be of a conscious (albeit ingrained) choice, as the strength of the conscious evolutionary advances for "selflessness" (being an adjunct of one's conscience); may deliberately override any instinctive fears of pending personal imminent harm. And in the case of a stranger, such heroism although frequently dismissed (by those same strangers themselves) as being nothing but a spur of the moment "instinctive reaction;" such conscious acts however, typically involve a high degree of (albeit ingrained and spontaneous) conscious choice. Either way, if it is true that the selfless act was itself a purely instinctive act, then this is a testament to the successful evolution of the concept of "selflessness" as being a direct extension of consciousness. Or, in the alternative, if a high degree of conscious choice was involved in choosing to put one's self at risk, then that too is a testament to humanity's evolved conscious ability to actually choose a different (genetic) course of action (by overriding ones instinctive fears) for the future benefit of others (again human or otherwise). And as a point of reference, reported cases abound where animals such as dogs, and mammals such as dolphins; are well documented to have put themselves in harm's way to protect humans in peril. Either way; the character trait of "selflessness" can become more and more a part of who we are (through natural selection); or, it can become a question of selfless intellectual choice, where as a question of "mind over matter;" within an individual (or other species) can also choose to put humanity first. But either way; "selflessness" (and its adjunct, "courage") are both advancements (genetically allowed) now within the cerebral brain (as they both either exist somehow as being consciously "ingrained," or they exist as a matter of conscious "choice"), but either way, both evolutionary advancements, are real, ... and (thankfully) both have proven themselves through human (and other wise) repetition; to be here to stay.

Such are the positive extensions built upon consciousness and conscience, whereby selflessness and courage are positive attributes because they promote and protect life and therefore can logically be construed as good for the overall evolutionary development of life. And it is here therefore that we find our best objective definition of what "good," actually is. And that is: "good," is that which is most consistent with the most progressive evolutionary developments that allow for the greatest growth and survival of life with its current accomplishments and advancements achieved thus far. And why is this, the best definition of "good;" because this is where evolution has brought us thus far. And humanity, in its current state is the latest achievement of this earth's conscious evolution. And mankind's conscience (with all of its latest advancements and development of character attributes) is the latest of what has been consciously accomplished (again) thus far. So what at first blush may appear to be selfish, we can claim that because we are at the apex of evolution (in that we claim for ourselves to be the single voice of earthly consciousness) we can declare, solely out of our shear ability to do so; that: that which is good for all life, as represented by "humanity" is by definition itself; "good." And although this looks paradoxically selfish, it is not necessarily so; because as humanity exists with our most evolved and advanced form of consciousness and conscience to preserve and protect all life, our ultimate achievement as a conscious species must therefore include our most evolved consciousness which includes compassion, empathy, generosity, kindness, selflessness, courage, sacrifice, love, intellect, curiosity, logic, truth, deductive reasoning, fairness and justice, to name just a few. All of these attributes being "good," because they represent that which evolution itself created through consciousness; and which evolution (now turbo charged through consciousness) can continue to use to nourish, protect and build upon, for the continuation and benefit of all life.

Conversely, "evil" is by definition the antithesis of creation and the evolution of life: it is the reality of senseless waste, death and destruction of life, (and yes that includes) any life (or its attributes). And it is here with this definition that we tend to misconstrue the nature of evil because we tend to visualize it with a consciousness that it doesn't necessarily have. Our natural instinct when we usually think about evil is to ascribe to it that which we believe is the evil consciousness and motives of the actor who is perpetrating the evil act (that being the destructive act) that has either already occurred, or is about to occur. And although in most cases the intent and motivations of those who perform acts of destruction; is exactly where we should be concentrating our urgent attention in order to combat the immediate destructive act at hand; as a matter of definition however, defining what "evil" actually is, is more accurately served if we step back and look beyond the particular actor (force or thing) and abstract the concept of "evil" so that our definition may be as all encompassing as possible. And it is with this in mind however, that we then recognize that "evil" in its most sterile definition, may be almost impossible to quantify because in its essence it may not only be a thing without form, but unlike symmetry and kindness (which are positives) without form, that through the power of creation, can become so real as to become themselves measurable. But conversely, to try to describe "evil" is to try to prove a triple negative; that is, based on (first) an existence of destruction (which is itself a negative); which then presents itself as the opposite of creation (which is a second negative), and then (thirdly) it assumes the immediate gratification, of the personal negative motives that (through a strong conflict of interest) for which that perpetrator then craves satisfaction. Such cravings as jealousy, envy, greed, vanity, ego, selfishness, apathy, arrogance and: this is just to name a few; to activate such destruction under the veil of such negative motivations is only to give a name to the resultant destruction, by …pulling the trigger as it were. So where good is aligned with creation, for evil to materialize, there needs to be destruction with the consequence of pain and hardship for evil to become identifiable. But regardless of what it is called; again it is the devastation and hardship which follows which is the measurement of the evil; because (from an evolutionary perspective) it is the loss that we must endure, and it is the sense of loss however which will bring about the evolutionary need for change, in the hope of securing future prevention.

And this triple negative then creates however its own paradox, making evil appear (despite its negative complexity) as an actual "power source" through the conscious allure of the immediate power gratification through the false accomplishment of destruction, which exists only as that which erases any creation thus far achieved. Evil therefore becomes real as a sensation because of the devastation realized through the destruction it inflicts. The paradox however exists because the loss it imposes, when brought about by human design, masquerades as an affirmative power for the fleeting sensation of gratification for a negative emotion; but destruction itself is only an annihilation of creation, which in turn (power as a destructive force) is an oxymoron itself. So what consciousness has achieved through evolution with the positive attributes and the characteristics that make life worth living, humanity is still left with the original attributes of self preservation for immediate and initial physical existence which evolution originally needed to create, to get us this far. It was the original reality where the earliest attributes prioritized "self" first, where the long term survival of the group (within the species) would be second; and the needs still for an even greater collective was/and would be left…to be (later considered) and possibly determined.

And it is here where we have to worry about the varying levels of conscious intellect within the various groups of mankind, where the selfishly inclined, can hijack the consciousness of others (within any human collective) through the conscious devices of deception and propaganda to destroy the "free will" of the trusting and the "ideologically dependent" (manipulated) masses. Also, it is here that we best similarly see why we have been so quick to attribute the conscious motivations of various selfish actors to "define the evil" of destruction that was employed, such as when we describe a fatal gunshot as a "jealous shooting." It is the act of senseless destruction however that is first (without more), an evil unto itself. It is only when the evil power of destruction is adopted by an actor, that the motivation of a negative attribute, such as jealousy, envy, greed or any of the others; it is only then, when the combination of destruction and motive, that the two become one; that the two become identified as (a particularized) form of "motivated evil." Experiencing, therefore; jealousy, envy or even moments of greed by themselves, are not demonstrations of evil (rather it might best be said, that to experience such sensations is "to be only human"), but when such negative extensions of our consciousness are allowed, through the breakdown of our intellect, to become obsessive, or all consuming; and then to act on such manifestations to engage in the act of destruction to the detriment of others, then such actions and intent do themselves become evil; even when the resultant harm is neither permanent nor necessarily long lasting.

Coming Full Circle: Consciousness, Gives Us The Right To Choose

So it is the understanding of what evil actually is that becomes vital. It is the distinction that it is the act of senseless destruction which is by definition "evil;" it is the motivated decision to invoke, or provoke destruction, either deliberately or through indifference which makes a negative motivation, (once acted upon) a particular form of evil. And at this point it is important to point out the fact that destruction itself with the advent of consciousness must also include the destruction of all things both material and that which exists without form. It includes the destruction of other such worldly things (or attributes) of creation that exist without form such as the destruction of trust, esteem, self worth, responsibility, confidence, and again so much more. Why, because from an evolutionary point of view; it is the pain, suffering and despair that result from such destruction that matters most (to possibly instigate evolutionary change); not the insignificant pettiness of a senseless perpetrator. Why, because as a collective if we concentrate solely on those who choose or cave into their negative attributes of existence, then when significant loss is experienced through destruction, where there is no intent or blameworthy perpetrator, for such things as death that results from child birth, famine or illness; such losses then run the hazard of being perceived as more comprehensible because it can be deemed to be the result of a (mere) "tragedy;" and therefore it can be understood or compartmentalized as being something that: "just happened/or happens." But such a compartmentalization of loss is however completely inconsistent with the purpose of the evolution of life, as evolution has brought the creation of consciousness thus far. Evolution created consciousness with its additional positive attributes to prioritize life and advance it beyond what evolution has itself been capable of doing; thus far. And we know this, because this is exactly what evolution has itself actually already accomplished by bringing us this far.

Consciousness, through the power of evolution /or creation, with the advanced attributes of conscience, comprehension, compassion, empathy, appreciation, responsibility, kindness and hope (and more); all have worked in the pursuit of a further evolving "consciousness," and therefore such an advanced consciousness cannot then rationalize (due to our additional attributes of intellect, perfected logic and deductive reasoning) to prioritize the value of life from a perspective of compartmentalization of understandable (innocent) losses. Such an understanding is completely inconsistent with all that evolution itself has created. Certainly, preventing those losses that can be prevented because they are brought about by any other human's negative intent, must be a top priority for all of us, simply because they are preventable; but such efforts should not be at the expense of us downgrading our "appreciation" of the lives lost due to other circumstances. So if we are to honor life at its utmost, then the senseless destruction of life must be understood as an evil unto itself, regardless of any intent behind it; simply because it was/and is senseless. And if we restrict the concept of evil to only that destruction that has conscious intent than we separate our losses in a way that they can never be perceived as equal. So as a matter of definition and appreciation for all lives senselessly lost; "evil" is the senseless destruction of life. And the conscious senseless taking of life therefore; is not any more or less evil, it's still the worst kind of evil, it's just more an identifiable evil as it is a representation of a failed (evolutionary) intellect of those who are responsible for the losses which deliberately (or negligently) did/or are about to, occur.

So famine, disease, thirst and all other evils that cause senseless pain, suffering and death; such things (from an evolutionary perspective) are all just as evil as any murder; it's just that with the taking of life through murder, we as a conscious societal/collective, have someone immediately at hand to hold conceptually accountable as a matter of internal/tribal (collective) responsibility. But the fact that famine, disease, thirst, birth defects, and all other forms of evil; are such things that cause suffering; the fact that such things are taking place beyond our conceptual walls of responsibility; does not mean that such pain and suffering and death are not also actually taking place, it just means that as a matter of economy they are not within our chosen tribal (conceptual) assumption of responsibility. So the choice of what we do, in the battle against hardship and suffering; the amount of responsibility we assume, is consciously ours to make; and we make these decisions based on conceptual groupings that pivot our conceptions of "self" and need, against our collective comprehension of (choice) for an ever larger, inclusion of humanity.

So senseless destruction is evil and it creates evil, and when we choose to embrace destruction, because of the false sense of power that it imitates and the real losses that it actually creates; then we choose to give life to evil by imposing suffering on others through destruction. So, yes individual humans and humanity as a whole can choose to be evil but we must understand that we make such decisions both through our actions, and through our neglect and inadvertence, as well. So in the end, senseless destruction is evil no matter where it comes from: and as for the impact that evil has on our evolution, well first we have to consciously choose to recognize it for what it is, both within ourselves individually and as a group, collectively; and then we have to evolve beyond it; and that in the end … as they say: is "evolution."

CHAPTER SIX:
EVOLUTION'S ASPIRATION

In the beginning, we posed the question what is the purpose of life, and we quickly came to the realization that to answer this question we needed to make sure that we had to have a comprehensive "understanding for what life is" (or could be) before we could address the follow up question: of "what is the purpose of life." And although we may have only scratched the surface of what life is or could be; our enquiry, though vastly incomplete, has not been now without its revelations. And what we have discovered is that life does have a purpose, and that purpose is to (at least) arrive, survive, evolve, achieve consciousness and then, continue to perfect that consciousness. And how can we know that this is the purpose of life, because, it has already happened.

Consequently, the purpose of life is to give our galaxy consciousness because that is what our galaxy has achieved for itself, thus far. Our galaxy exists, plain and simple. And our solar system came to be, because of our galaxy, and our earth grew out of the creation of our solar system. And all life on earth came to be after the cooling of our planet. From there, all the species of the earth slowly evolved into being, and with being: came consciousness. The question of whether this could all come to be, by random selection or occurrence; does not mean that the living entities so far achieved, are therefore without purpose. What it does mean, is that all that: "is," has arrived as a result of a driving force/or forces, and that this/or these driving force(s) turned inorganic gas and matter, eventually into organic matter with consciousness; and that between these driving force(s) and the physical matter that was so engaged, that these forces and materials had the ability to create life, and subsequently later, life with consciousness. Thus such driving force(s) was/ is/ are or ought to be, presumptively evolutionary; if not also undeniably progressive? And it matters not, as has been argued here, that such an evolutionary progression denotes signs of some sort of indiscernible "cognitive properties" (on this we can simply choose to disagree); what matters (again) is: that, what has happened; has actually already occurred; and if there is no "cognition," then the argument that our galaxy brought us here by pure chance, this makes all earthly consciousness a pure miracle, which should mean that our appreciation of all life, should therefore be even more revered. And, in the event that all creation has come about by chance; then because we are now conscious (with the latest advancements of conscience and everything else), then we have an even greater responsibility to "everything else" to give the benefits of that advanced form of consciousness to "everything else." And yes that means that this responsibility extends from the abilities (and rights) of the individual to humanity as a whole, thereby recognizing that these powers therefore rightly belong (with their origin being in the galaxy) equally, to the individual, to humanity, our posterity; and to everything else.

So: who was/could have been/ or is, the next Marie Curie or Albert Einstein; and how many of humanity's best have been lost to child birth? Lost to disease? Famine? Been executed? Been enslaved? Imprisoned? Subjugated and/or oppressed? These are all the questions that defy evolution because senseless loss of life (or the things that defeat it), are themselves things that are anti-evolutionary.

Taken as a whole, evolution has had an undeniable trajectory. The very creation of life itself with its many varied species, with their equally varied abilities and attributes; and the potential in each case to continue to become "more" than what has been achieved thus far, means that evolution and life are one in the same, acting as a spiraling force of ascension with a symbiotic trajectory. And positive things that exist, without form, such as: symmetry, harmony, truth, logic, precision, reasoning etc.; all take on a sense of tangible reality through the conscious (material) creation of that which can be actually physically seen touched and appreciated. And such things stand as consequent proof of their own existence, through the manifestation of such measurable things as art, music, science, engineering, architecture, etc.; while, all-the-while this spiraling force of evolution and creation perpetuates (through consciousness) other such things (without form) attributes such as love, compassion, kindness, generosity, fairness, empathy, hope, justice and more; all these things becoming real, in order to ensure that life itself will always be worth something more.

And herein lies with this (our current analysis) possibly the most powerful revelation of all; which is: that with the realization that there exists an earthly/ galactic aspiration for not only consciousness, but an ever perfecting sense of consciousness; a pursuit that means that we should recognize a new realm of quantitative reasoning and science for the appreciation of these and other evolving attributes. A science that understands and theorizes that what was once thought to be nothing more than emotional niceties/or aberrations, that such attributes were/and are in fact conscious extensions of our consciousness for a reason. It is the quantitative understanding that such attributes are in fact evolutionary advancements which were/ and are the precursors to still even higher levels of appreciation for all things. And it is the realization that the (currently unknowable/cognitive) driving forces of our galaxy; are leading us, to "consciously," perpetuate further attributes (such as a greater sense of overall 'responsibility') for the betterment of all life and humanity. And how do we know this, because this is exactly the evolutionary place upon which we have already arrived. We have evolved now to the point where we realize that attributes such as kindness stimulates the physiological healing and wellness through the natural creation of hormones and other stimulants such as serotonin, endorphins and Oxytocin. So for conscious attributes, what were once thought to be signs of weakness during more primitive and ruthless times; the development of such attributes may have been actually experiencing evolutionary progressions where the more compassionate side of humanity was aspiring to champion its cause through the process of enlightenment.

So historically, from the primitive man at the time of the Neanderthals, through the enslavement of the Pharaohs, to the cruelty and barbarism of the Vikings, to the omnipotence of Royalty, to where we stand now, with the pretext of Constitutional democracies; the compassionate attributes of mankind have endured and in some instances, may now even be said to have been institutionally codified. And what this means is that the compassionate and the aspiring (equilibrium) attributes of mankind, are the latest evolutionary progressions of human consciousness. And the attributes that facilitate the "good will" and "connection" for all of mankind are therefore themselves to be seen as representative of the evolutionary course, for all of mankind. And this means that the compassionate attributes of mankind (that were once viewed in earlier ego centric power confrontations as weaknesses) are now to be distinguished from this construct and are now best seen as part of the evolutionary cure separating us from the hardships and disparities of physical existence. So now, from a life perspective, it is best seen that a child tending to a wounded bird is itself the facilitation of the latest extension of the continuum of quantum physics as the compassionate attributes of consciousness are being further stimulated in both the child and the attended species alike. And what we need to do, through our own evolution, as observers; is to figure out just how to try to explain this phenomenon to the two of them, as best we can. Because as discussed earlier, in aspiring to do the teaching, the aggregate itself will become the teacher.

And what all of this means, is that: as we look upon ourselves as individuals and we then turn and look upon others and their actions, both past and present; it becomes ascertainable as to whether our actions individually or collectively, have been in sync with this overall trajectory for the harmonization of life with the aspirations and efforts to date, of our aspiring galaxy. Why, because as individuals, we and the other species of this earth are the consciousness of our aspiring galaxy (currently) in the aggregate. So the question then becomes, have we humans failed to consciously evolve both as individuals and through our chosen collectives, to evolve as life forms; or is our failure to be measured by our preoccupation with the immediate superficial distractions and addictions towards "one's" self as showcased through the pursuit of such personal seductions as: power, greed, vanity and ego centrism. Because in the end, since we are all destined to leave this earth; if in the end we have inspired no one (or nothing); than an individual life spent on self gratification, in the hereafter, will have little understanding within the greater synchronicity of the life forces operating within a higher interconnected "universal" reality. A reality, which from its very beginning, has endeavored to do its part.

…But even for something as "special" as this earth, with all that it has to offer, time eternal; … still waits for no one.

www.ingramcontent.com/pod-product-compliance
Lightning Source LLC
Chambersburg PA
CBHW031922240526
45464CB00021B/635